Molecular Biology: Current Innovations and Future Trends.
Part 1.

The ***Current Innovations in Molecular Biology*** series

Volume		Publication Date
1.	Molecular Biology: Current Innovations and Future Trends. Part 1.	April 1995
2.	Molecular Biology: Current Innovations and Future Trends. Part 2.	October 1995
3.	Internet for the Molecular Biologist.	March 1996

For further information on these books please contact:

Horizon Scientific Press
P.O. Box 1
Wymondham
Norfolk
NR18 0EH
U.K.

FAX +44-1953-603068

MOLECULAR BIOLOGY

CURRENT INNOVATIONS AND FUTURE TRENDS

Part 1

Volume 1 in the ***Current Innovations in Molecular Biology*** series

Edited by

Annette M. Griffin
Institute of Food Research, Norwich Research Park, Colney, Norwich NR4 7UA, England

Hugh G. Griffin
Institute of Food Research, Norwich Research Park, Colney, Norwich NR4 7UA, England

horizon scientific press

Horizon Scientific Press
P.O. Box 1
Wymondham
Norfolk NR18 0EH
England

British Library Cataloguing-in-Publication Data

Molecular Biology: Current Innovations and Future Trends. -
(Current Innovations in Molecular Biology Series; Vol. 1)
I. Griffin, Annette M. II. Griffin, Hugh G. III. Series
574.88

ISBN: 1 898486 01 8

Printed and bound in Great Britain by
Biddles Ltd, Guildford and King's Lynn

Preface

Molecular biology is a fast-evolving science. Ideas, concepts, and technology advance at a lightening pace and it is vital for molecular biologists to keep abreast of current developments. Many excellent textbooks exist that contain protocols for specific techniques. However, it was felt that there was a need for a series of books to keep scientists informed of recent innovations in both the theory and practice of molecular biology. The new series ***Current Innovations in Molecular Biology*** will cover current developments and advances relevant to all areas of molecular biology. Each volume will contain comprehensive reviews on the theory, methodology, and applications of the most recent and widely used techniques in molecular biology.

Volume 1, *Molecular Biology: Current Innovations and Future Trends Part 1*, covers the important topics of PCR, cycle sequencing, gel electrophoresis, PFGE, DNA purification, automated DNA hybridization, subtractive hybridization, and oligoribonucleotides. The book also describes the applications of capillary isoelectric focusing and mass spectrometry in molecular biology.

Volume 2, *Molecular Biology: Current Innovations and Future Trends Part 2*, covers automated sequencing, phylogenetics, electroporation, non-radioactive labelling, magnetic beads, peptide nucleic acids, antisense technology, protein sequencing and, in addition, describes the applications of HPLC and NMR in molecular biology.

Volume 3, *Internet for the Molecular Biologist*, aims to demystify the information superhighway by answering questions such as: What is it? How can I do it? What can I use it for? How useful is it for molecular biology? The book is written specifically for the molecular biologist and contains a wealth of information and practical instruction vital to the researcher of today.

Further volumes on other important areas of molecular biology are planned for the future.

It is hoped that the current volume fulfils its ambition to provide researchers with an overview of recent advances in the theory and practice of molecular biology. Each chapter provides a concise yet comprehensive review of the most recent innovations in a specific technology. Most chapters also provide step-by-step protocols and, in addition, discuss future trends for technological improvement. We hope that *Molecular Biology: Current Innovations and Future Trends* provides an insight into the newest areas of molecular biology and presents a wide range of new techniques for even the most experienced researcher.

The editors express their gratitude and appreciation to all those associated with the production of this volume and in particular to the contributing authors.

Annette M. Griffin
Hugh G. Griffin

Contents

Contributors

John C. Bauer
Stratagene Cloning Systems
11011 N. Torrey Pines Road
La Jolla
CA 92037-1029 USA

Stephan Beck
Imperial Cancer Research Fund
London
WC2A 3PX UK

Gina L. Costa
Stratagene Cloning Systems
11011 N. Torrey Pines Road
La Jolla
CA 92037-1029 USA

Roel Funke
Plant Molecular Genetics
University of Tennessee
Knoxville
TN 37901-1071 USA

Tim Gackstetter
Stratagene Cloning Systems
11011 N. Torrey Pines Road
La Jolla
CA 92037-1029 USA

Christian E. Gruber
Life Technologies Inc.
8717 Grovemont Circle
Gaithersburg
MD 20898 USA

Ivo Gut
Imperial Cancer Research Fund
London
WC2A 3PX UK

Paul N. Hengen
National Cancer Institute
Frederick
MD 21702-1201 USA

Alexander Kolchinsky
Plant Molecular Genetics
University of Tennessee
Knoxville
TN 37901-1071 USA

Branko Kozulić
Guest Elchrom Scientific AG
Gewerbestrasse 10
CH-6330 Cham Switzerland

Keith A. Kretz
Stratagene Cloning Systems
11011 N. Torrey Pines Road
La Jolla
CA 92037-1029 USA

Wu-Bo Li
Life Technologies Inc.
8717 Grovemont Circle
Gaithersburg
MD 20898 USA

Tom Pritchett
Beckman Instruments
2500 Harbor Boulevard
Fullerton
CA 92634 USA

Ravi Vinayak
Applied Biosystems Division
Perkin Elmer
850 Lincoln Centre Drive
Foster City
CA 94404 USA

Michael Weiner
Stratagene Cloning Systems
11011 N. Torrey Pines Road
La Jolla
CA 92037-1029 USA

From: *Molecular Biology: Current Innovations and Future Trends.*
ISBN 1-898486-01-8 ©1995 Horizon Scientific Press, Wymondham, U.K.

1

RECENT ADVANCES IN PCR METHODOLOGY

Michael P. Weiner, Tim Gackstetter, Gina L. Costa
John C. Bauer, and Keith A. Kretz

Abstract

Protocols are presented for several PCR methods. These include *Pfu* DNA polymerase polishing to increase PCR cloning efficiency, blunt-ended PCR cloning, site-directed mutagenesis and colony PCR. The objective of all of these methods when used with PCR is to decrease both the time and the effort involved in routine laboratory procedures. A discussion of laboratory equipment and supplies, including thermocyclers and imaging equipment is included. We also present a brief discussion of the different polymerases used for PCR, including mixed polymerases used for PCR of long regions of DNA.

Introduction

The Polymerase Chain Reaction (PCR) is a powerful technique which can selectively amplify a specific segment of DNA out of a complex of nucleic acids. In this chapter we will discuss some of the basic materials and techniques needed for successfully performing PCR. Protocols will be provided for troubleshooting the PCR using the different polymerases, cloning PCR-derived DNA fragments, screening of recombinant colonies (colony PCR), PCR-based DNA sequencing and site-directed mutagenesis of DNA using PCR.

Laboratory Equipment and Supplies

Instruments to support research utilizing PCR have been evolving as rapidly as the applications for PCR. Present day instruments include thermal cyclers, ultraviolet light sterilizers, and improved methods of gel imaging and analysis.

Thermal Cyclers

All thermal cycling instruments take one of two approaches to cycling the temperature of the samples. Either samples are placed into a single block and the temperature of that block is changed or the samples are moved between different temperature blocks. The advantages and disadvantages of each approach will be explored as well as the evolution of the instruments.

The first PCR experiments were performed by moving samples between water baths at different temperatures by hand. Although the temperature uniformity between samples was good, the temperature accuracy was not. Also, the potential for human error was high since the researcher was tasked with moving samples every few minutes over the course of a few hours.

The first instruments consisted of a metal block with wells where the samples were placed. The block had internal channels to circulate heated and cooling fluids to change the temperature of the block as programmed. Although the task of cycling samples between temperatures had been simplified, the instrument was very expensive and the temperature of the sample wells varied across the block. Well-to-well temperature variation has been shown to affect PCR amplification efficiency (1).

During this time, other researchers took a more obvious approach to automating the process whereby samples were moved between water baths. An instrument used for embedding tissue samples, called a Histokinette, was modified for PCR protocols (2). The modified Histokinette used a robotic arm to move samples according to a programmed routine between appropriately temperatured water baths. This instrument provided a low variation of temperature between the samples. However, since the modified instrument was not available commercially, it was up to each laboratory to make the changes to its own Histokinette.

Fortunately, instruments for temperature cycling have continued to evolve. Initially, in the single block cycler, a heating element was added directly to the bottom of the block. The block was cooled by running tap water through internal fluid channels in the block. Although this reduced the cost of the instrument, well-to-well temperature variation was still a problem in addition to having to contend with leaking valves and wasted water. To overcome this problem, some instruments were built using solid state thermal elements, called Peltier devices. These devices work on the principle that a current flowing between different metals will draw thermal energy away from one metal and into the other causing one side to become cool and the other to get hot. When the current is reversed, the heating and cooling effect is also reversed enabling the hot side to become cold. The Peltier device remedied the problem of messy fluids and valves. However, due to the thermal expansion and contraction stresses placed on the Peltier devices, they suffered a high failure rate. This is presently being improved upon by two different methods. Either multiple smaller sized Peltier devices are used to reduce the thermal stress in each device, or grooves are cut into a single larger device to better absorb the stresses.

Unfortunately, an inherent problem still remains when trying to change the temperature of a metal block. The two most desirable traits of a thermal cycler are the rapid cycling times and the temperature uniformity between every sample. However, as the mass of the block is reduced to decrease the time required to change the block's temperature, the ability to maintain a uniform temperature between every well becomes more difficult. Some laboratories alleviate this problem by avoiding the use of the outer rows of wells in their thermal cycling instruments.

Alternate approaches to a metal block include machines that cycle samples in a chamber using heated air. One such instrument uses custom positive displacement pipette tips (3). The advantages of such a system are small volumes (0 to 25 µl), reduced cycling times, and reagent costs are decreased. However, the disadvantages are additional handling required to seal and then cut open the custom tips, and the temperature accuracy of the hot air is less than that of the metal block system. Another instrument, available commercially is similar to an oven and has a rotating carousel (4). Although tubes, trays, or slides can be placed inside, it also exhibits decreased temperature accuracy.

While the instruments mentioned above have been evolving, the number of sample formats has also increased. Initially, samples were placed in 1.7 ml microcentrifuge tubes, although it was found that 50 to 100 µl of sample size was sufficient and instruments were introduced for 0.5 ml tubes. To help reduce the thermal mass of the tubes and decrease the cycling times, thin-wall 0.5 ml tubes were used. As it again became apparent that sample sizes could be reduced as well as reducing the cost of each reaction by decreasing the amount of reagents, particularly the DNA polymerase and the synthetic primers, more and more instruments were introduced to accommodate the 0.2 ml thin wall tubes. Most of these instruments can also be used with 96 well trays. Recently, *in situ* PCR techniques have been published where the PCR is performed using glass slides (5).

Sample Preparation and Sterilization

Since PCR will just as readily amplify contaminated DNA samples as pure ones, it has become apparent that care must be taken when preparing samples. One method of sterilization is to expose the labware and some of the reagents to ultraviolet (UV) light (6, 7). The UV light probably functions by causing inter- and intra-strand pyrimidine dimerization through which the polymerases are incapable of synthesizing complementary DNA. An instrument (Stratalinker, Stratagene; La Jolla, CA), originally introduced to crosslink nucleic acids onto membranes, can also be used for PCR sterilization (7). Samples can be placed in the instrument and exposed to UV light either for a set amount of time or for a specified amount of total energy (6, 7). To inactivate contaminating template, it is recommended that 200-300 mJ/cm^2 be used on the reaction buffer, and approximately 5-6 times that amount of UV irradiation on solutions containing primers and nucleotides (7).

Uracil DNA glycosylase (UDG: uracil-N-glycosylase) removes residues from the sugar moiety of single and double-stranded DNA. The resulting abasic sites are cleaved at elevated temperatures. When used in conjunction with dUTP, the UDG enzyme can be used to remove carryover contamination in PCR reactions (8).

Imaging

The most common method of analyzing PCR products is by gel electrophoresis. The sample, along with standard markers, is loaded onto an agarose gel, and subsequently stained using ethidium bromide. The DNA bands are then observed using a UV light source. In the past, an instant-film camera was used to take a photograph of the gel. Recently, still video systems have been introduced as an alternative to the film camera. Still video systems use a camera with a charge coupled device (CCD) to capture an image which is then digitized and stored in a computer. The image can be analyzed using densitometry software, archived onto floppy disks or hard drives, sent to a slide maker for presentations, or printed on a thermal printer. The digital image is easily transferred electronically to colleagues and to manuscripts for publication. Thermal prints are one tenth the cost of Polaroid prints. As an added benefit, the need to take multiple Polaroid prints to optimize exposure is eliminated since the image that is displayed on a monitor can be modified before it is sent to the thermal printer.

Thermostable Polymerases used for PCR

The isolation of a thermostable DNA polymerase from *Thermus aquaticus* (*Taq* DNA polymerase) allowed PCR to become an easily practiced technique (9). This overcame the necessity of adding fresh DNA polymerase after each denaturation cycle and allowed the annealing and extension reactions to be performed at a stringent temperature. Since the advent of the use of *Taq* DNA polymerase, a large variety of thermostable DNA polymerases have been developed for use in PCR (Table 1). Many of these new DNA polymerases have properties very similar to *Taq* DNA polymerase, while some have properties which make them significantly different. A list of frequently used enzymes and their properties is provided in Table 1.

Table 1. Enzymes commonly used in PCR

		Selected Activities		
Enzyme (reference)	Source organism	5' to 3' exonuclease	3' to 5' exonuclease	3' extendase[1]
Taq (10)	*Thermus aquaticus*	yes	no	yes
Stoffel[2] (10)	*Thermus aquaticus*	no	no	unk
Tli (12)	*Thermococcus litoralis*	no	yes	unk
Pfu (14)	*Pyrococcus furiosus*	no	yes	no
Tma	*Thermotoga maritima*	yes	yes	unk
Tth (16)	*Thermus thermophilus*	yes	no	unk
Tub (17)	*Thermus ubiquitus*	yes	no	unk

[1] 3' mononucleotide terminal deoxynucleotidyl-transferase activity: yes, contains extendase activity; no, no measurable activity; unk, extendase activity not known.

[2] Amino terminal deletion of *Taq* DNA polymerase.

Taq DNA Polymerase

Taq DNA polymerase is still the most widely used and best characterized enzyme for PCR. This 94-kDa enzyme has no 3' to 5' exonuclease activity (proofreading activity) associated with it's 5' to 3' exonuclease activity and DNA dependent-DNA polymerase activities. The DNA polymerase has a maximal activity at 75°-80 °C in 2-4 mM $MgCl_2$ and 10-55 mM KCl with a half-life of 45-50 minutes at 95 °C (10). The 5' to 3' exonuclease activity is unusual in that it is not active on single-stranded DNA but rather catalyzes the digestion of DNA annealed to the template downstream of an active polymerase molecule (10). Various *Taq* DNA polymerase clones devoid of the 5' to 3' exonuclease activity have also been produced and characterized (10, 11). The

Stoffel fragment, a 288 amino acid deletion, has been shown to be slightly more thermostable than full length *Taq* DNA polymerase, optimally active over a wider range of Mg^{+2} concentrations and less processive (10). KlenTaq, a 238 amino acid deletion, has been reported to insert mutations at a two-fold lower mutation rate than full length *Taq* (11).

Other Thermostable Polymerases

Thermostable DNA polymerases with proofreading capabilities have now been described from *Thermococcus litoralis* (*Tli*, 12, 13), *Pyrococcus furiosus* (*Pfu*, 14) and *Thermotoga maritima* (*Tma*, 15). The polymerases from *Tli* and *Pfu* have been shown to have a higher fidelity than non-proofreading enzymes such as *Taq* (13). One study has demonstrated a ten-fold improvement in the error-rate measured for *Pfu* DNA polymerase (1.6×10^{-6}) when compared to *Taq* DNA polymerase (2×10^{-5}, see also reference 14). These findings are significant for PCR techniques which require high-fidelity DNA synthesis, including the direct cloning of PCR amplification products, PCR-based procedures for high-efficiency double-stranded mutagenesis and amplification techniques designed to detect specific point mutations.

Other DNA polymerases that have been successfully used for PCR include *Tth* (16) and *Tfl* (17, 18).

Mixed Polymerases for Long PCR

Recent advances have extended the length of DNA fragments which can be amplified with the PCR method (19, 20, 21). Small amounts of thermostable, proofreading enzyme used in conjunction with either *Taq* (or *Tth*, 16) DNA polymerase allows the synthesis of 35-40 kb stretches of DNA from non-complex templates (lambda clones) and up to 20 kb from complex templates. It has been proposed that the non-proofreading enzyme (*Taq* or *Tth*) incorporates mismatched nucleotides at a sufficient rate to stall DNA synthesis of fragments larger than ~5-6 kb. The presence of a small amount of proofreading enzyme could then remove mismatched nucleotides allowing the extension of any previously stalled amplicon. The positive effect of mixed polymerases seems to be dependent on proofreading activity of one of the added polymerases. Substitution of a non-proofreading enzyme for the proofreading enzyme (ie. exo$^-$*Pfu* replacing wt-*Pfu*) does not result in the ability to synthesize these long PCR products (19).

In addition to the 3' to 5' proofreading, the extendase activity of some thermostable DNA polymerases may also be required for this long PCR. The reason for this may be the need for extending/polymerizing a nontemplate-directed base across an abasic site in the template DNA during PCR. Such abasic sites might arise from depurination or depyrimidation during pH and temperature changes during either the reaction itself or template preparation.

The optimal ratio of proofreading to non-proofreading enzyme appears to be quite broad. We have found that different ratios of *Taq*:*Pfu*, work well in a wide range of applications. It has also been demonstrated that the addition of glycerol and/or DMSO at various concentrations to the reaction mixture may increase the yield of long PCR products (20, 21). Additional evidence shows that a higher pH (8.8-9.0) and stronger buffering capacities can result in more successful long PCR reactions. The best reaction conditions, including buffer and additives, appear to be highly variable and should be optimized for every primer/template combination.

Specific Applications of PCR in Molecular Biology

Specific applications which support research utilizing PCR have been evolving at an extremely rapid rate. Present day applications include long-PCR, PCR polishing, cycle sequencing, site-directed mutagenesis and PCR cloning. For most applications dealing with PCR, commercial systems are available containing pre-tested reagents and protocols.

DNA Sequencing using PCR

The demonstration that *Taq* DNA polymerase could be used in a dideoxynucleotide terminator sequencing format opened the door for the development of cycle sequencing (22, 23). Cycle sequencing is the combination of temperature cycling found in PCR with dideoxynucleotide terminator sequencing as used in the standard dideoxynucleotide sequencing reaction. While cycle sequencing does not generate a logarithmic increase in yield, the linear amplification obtained in 20-30 rounds of temperature cycling significantly increases the signal and allows the use of much smaller amounts of DNA template. Repeated rounds of denaturation and primer annealing also allow a much wider range of DNA (cosmids, lambda DNA and PCR products) to be used as sequencing templates. A detailed analysis and review of DNA sequencing using PCR can be found in Chapter 2.

Cloning PCR-Generated DNA Fragments

Many methods exist with which to clone PCR-derived DNA fragments. These methods can be separated into those which require the addition of extra bases to the PCR primers, and those which do not require any added bases (reviewed in reference 24). The PCR cloning methods which necessitate primer modifications are shown in Figure 1. Group I includes methods for the incorporation of appropriate restriction enzyme target sequences on the PCR primers, and the UDG and ligation-independent cloning (LIC) protocols which generate 9-12 base single-stranded ends on the PCR fragment. Group II include the various T/A methods which rely on the terminal extendase activity of some DNA polymerases (25, 26), and also the blunt-end cloning methods which selectively clone blunt-ended molecules (27, 28). Group I methods routinely yield extremely high efficiencies, yet due to the ease and lack of expense, Group II methods are often preferred.

Some of the older polymerases commonly used for PCR have different and quite specific extendase activities (26). This characteristic will decrease cloning efficiencies associated with the T/A cloning methods. Therefore, unless it is known which base is preferentially extended onto the end of completed PCR molecules, blunt-ended cloning should be used (see protocol 2 below). Protocol 1 (PCR-insert polishing) should be

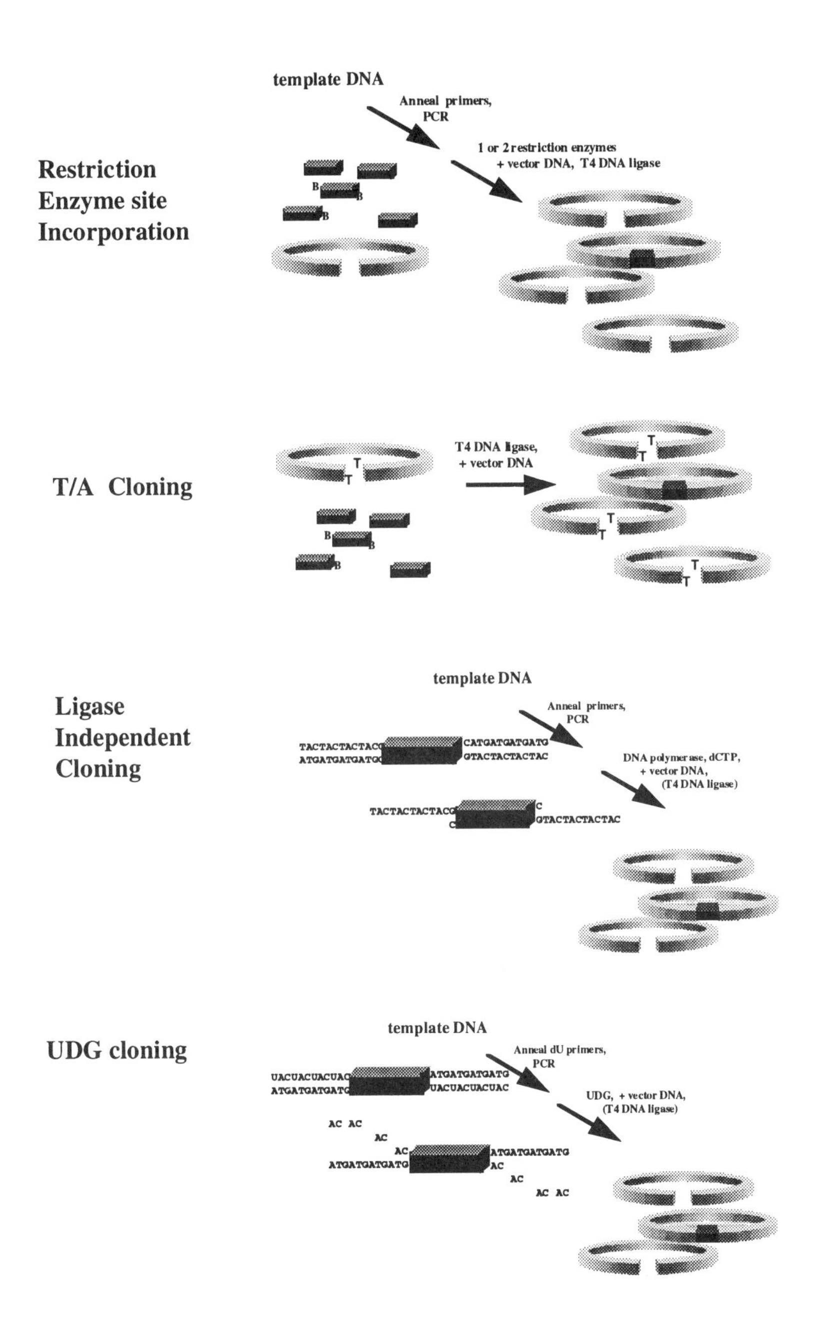

Figure 1. Methods for PCR cloning. (A) Restriction site incorporation in which appropriate sites are incorporated at the ends of the PCR primers; (B) T/A cloning which is dependent on the base-specific 3' template-independent mononucleotidyl-terminal transferase activity of some enzymes used for PCR; (C) Ligation independent cloning (LIC) which uses the 3' to 5' exonuclease activity of an appropriate polymerase to remove a predetermined number of bases from the 3' end of completed PCR fragments; and (D) UDG, which uses the enzyme uracil DNA glycosylase to remove PCR primers synthesized with modified (uridine) bases.

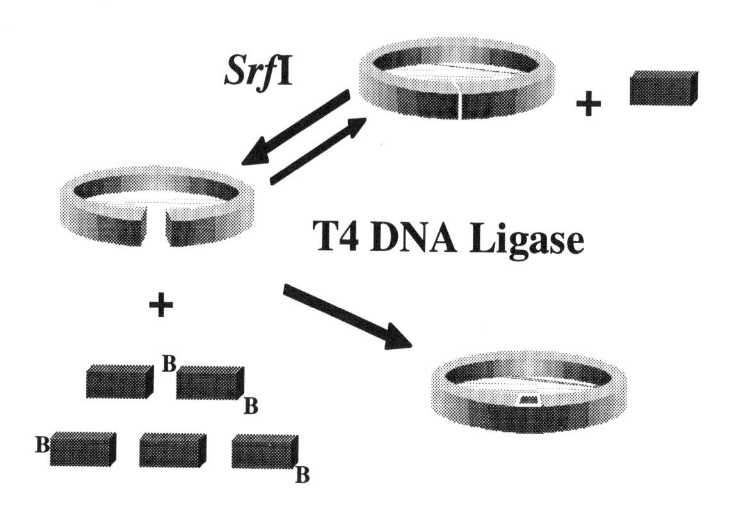

Figure 2. Method for blunt-ended PCR cloning. More efficient blunt-ended PCR cloning includes the addition of a restriction enzyme to the T4 DNA ligase reaction (in this example *Srf*I endonuclease) to regenerate the linearized vector during the ligation reaction.

used for removing extended bases on insert DNA or in cases where the extendase activity of the polymerase used to generate the PCR insert is not known (29, 30). *Pfu* DNA polymerase (Stratagene, La Jolla, CA) is preferentially used in this protocol because it has the ability to remove extended bases (by its 3' to 5' exonuclease activity) and is active only at increased (>65 °C) temperature (29). At the temperature used for ligation (22 °C), the polishing enzyme is inactive and therefore does not need to be removed prior to a ligation treatment.

Protocol 2, PCR cloning, increases the yield and efficiency of blunt-ended cloning by incorporating a unit excess of restriction enzyme (compared to the units of T4 DNA ligase, see Figure 2) in the ligation reaction (27, 28). The added restriction enzyme serves to keep an excess of linearized vector in solution by restriction of plasmids that religate intramolecularly. Intermolecular ligation with an insert destroys the endonuclease target site. In addition, a blue-white phenotypic color selection is used to differentiate molecules that contain an insert. Although any blunt-ended restriction enzyme for which the plasmid contains a single site can be used for this procedure, it is preferable to use a rare-target site enzyme. The commercially available restriction enzyme *Srf*I (31, Stratagene, La Jolla, CA) has been successfully used in our laboratory. It has the advantages of an octameric length of its target sequence (5'-GCCC|GGGC-3') and rareness in mammalian DNA (estimated at 1 in 10^5 bases). These characteristics make *Srf*I an ideal restriction enzyme for the restriction/ligation reaction because the recognition sequence should not occur in most PCR fragments.

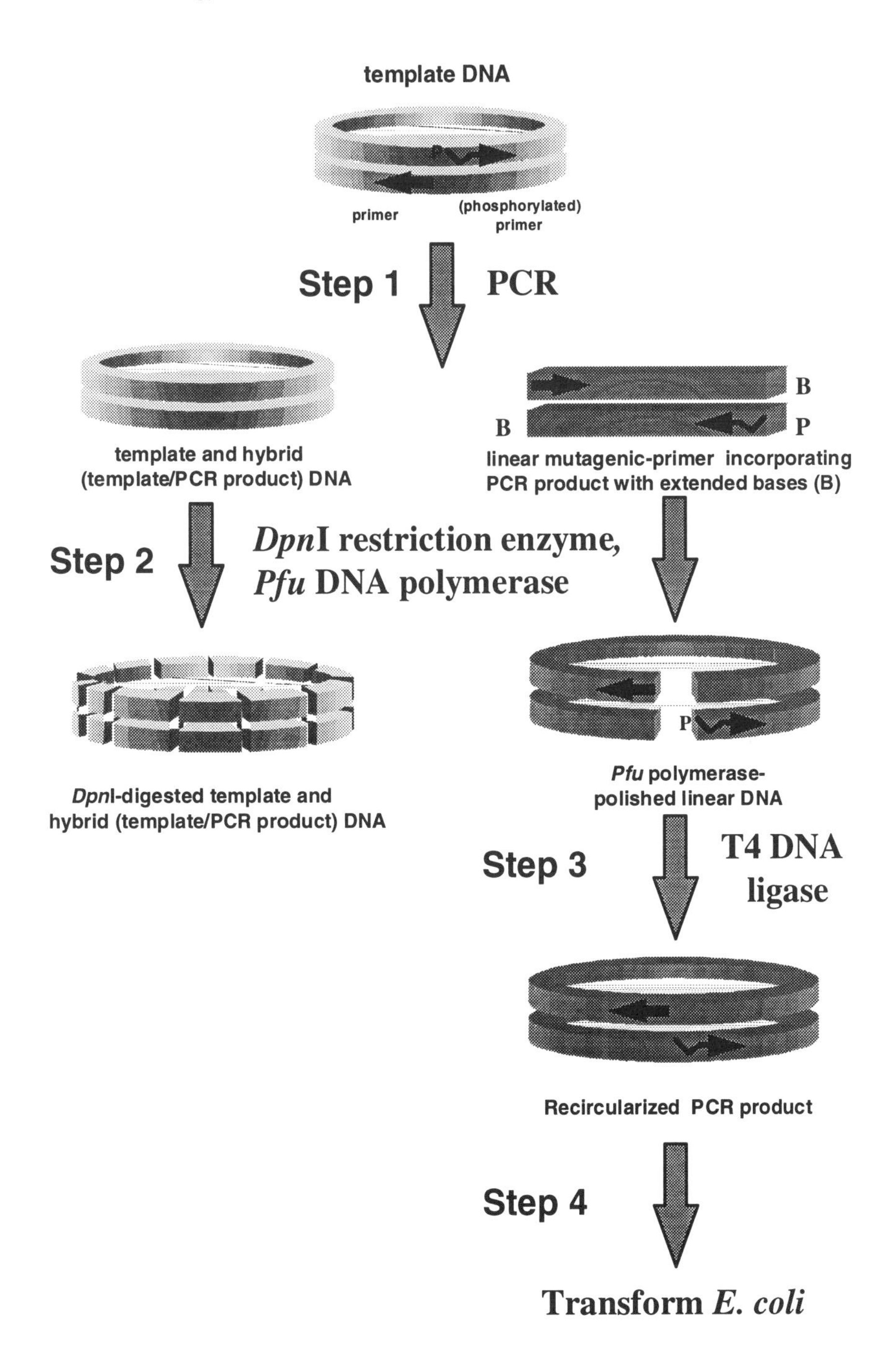

Figure 3. PCR based site-directed mutagenesis. *Taq* Extender is used in this method to aid in generating the complete template (step 1). The enzymes *Dpn*I endonuclease and *Pfu* DNA polymerase are used to restriction-digest parental molecules and polish the extended bases from the newly synthesized PCR fragments, respectively (step 2). T4 DNA ligase is used to recircularize the PCR fragments (step 3) prior to transformation (step 4).

Site-directed Mutagenesis using PCR

In vitro site-directed mutagenesis is an invaluable technique for studying protein structure-function relationships, gene expression and vector modification. Several methods have appeared in the literature, but many of these methods require single-stranded DNA as the template. The reason for this, historically, has been the need for separating the complementary strands to prevent reannealing. Use of PCR in site-directed mutagenesis accomplishes strand separation by using a denaturing step to separate the complementing strands and allowing efficient polymerization of the PCR primers. PCR site-directed methods thus allow site-specific mutations to be incorporated in virtually any double-stranded plasmid; eliminating the need for M13-based vectors or single-stranded rescue (32).

Several points should be mentioned concerning site-directed mutagenesis using PCR. First, it is often desirable to reduce the number of cycles during PCR when performing PCR-based site-directed mutagenesis to prevent clonal expansion of any (undesired) second-site mutations. Limited cycling which would result in reduced product yield, is offset by increasing the starting template concentration. Second, a selection must be used to reduce the number of parental molecules coming through the reaction. Third, in order to use a single PCR primer set, it is desirable to optimize the long PCR method described above and in references 12 and 13. And fourth, because of the extendase activity of some thermostable polymerases (25, 26) it is often necessary to incorporate an end-polishing step into the procedure prior to end-to-end ligation of the PCR-generated product containing the incorporated mutations in one or both PCR primers.

Protocol 3 is provided as a facile method for site-directed mutagenesis (see also Figure 3) and accomplishes the above desired features by the incorporation of the following steps: (i) increasing template concentration approximately 10^3-fold over conventional PCR conditions; (ii) reducing the number of cycles from 25-30 to 5-10; (iii) adding the restriction endonuclease *Dpn*I (recognition target sequence: 5'-G^{m6}ATC-3', where the A residue is methylated) to select against parental DNA (33, note: DNA isolated from almost all common strains of *E. coli* is Dam-methylated at the sequence 5'-GATC-3'); (iv) using *Taq* Extender (32, 34) in the PCR mix for increased reliability for PCR to 10 kb; (v) using *Pfu* DNA polymerase to polish the ends of the PCR product, and (vi) efficient intramolecular ligation in the presence of T4 DNA ligase.

Colony-screening using PCR

PCR screening of recombinant clones is a rapid and efficient method and results in a substantial time saving over conventional screening methods such as mini-preparation of DNA (35, 36). Protocol 4 is performed in only a few hours and can be used to determine the presence of an insert. The resultant PCR product can also be subjected to standard restriction enzyme analysis and DNA cycle sequencing.

Future Trends

Instrumentation

The future of thermal cyclers will include instruments with increased cycling speeds and better temperature uniformity, smaller sample size to lessen the cost per sample, and increased automated sample preparation, cycling and analysis.

Human Genome Project

PCR will continue to play an ever-expanding role in the project to sequence and understand the human genome. The cloning of DNA fragments has been simplified and accelerated by the powerful PCR process, which now makes it possible to generate an insert, clone and transform it into *E. coli* in a single day. To speed this process even further will require a collaborative effort of structural biophysicists, chemists, molecular biologists, and others. Better polymerases with the ability to extend the range and fidelity of PCR generated fragments will be available in the near future. New mapping and DNA sequencing techniques will be applied to, and developed for the human sequencing project.

Protocols

Protocol 1. PCR Insert Polishing

Insert polishing is used to remove extended bases from PCR-generated DNA fragments (29, 30). In general, this will result in an increased amount of blunt-ended DNA molecules available for the cloning reaction. PCR insert (10-500 ng) is added to a 10 µl reaction mix containing 1 µl of *Pfu* DNA polymerase (2.5 U/ µl, Stratagene), 1x *Pfu* polymerase buffer, and 1 mM dNTP (0.25 mM *each* nucleoside triphosphate). A mineral oil overlay is added and the reaction is incubated for 30 minutes at 72 °C. A 1-2 µl aliquot can be used in the PCR cloning protocol (below).

Protocol 2. PCR Cloning

Restriction-enzyme (*Srf* I, see reference 31) digested plasmid template DNA (approximately 10 ng) is added to a 10 µl reaction mix containing 0.5 mM ATP, 1x reaction buffer (0.1 M Potassium Acetate, 25 mM Tris-Acetate, pH 7.6, 10 mM Magnesium Acetate, 0.5 mM 2-mercaptoethanol), 2-4 µl PCR-generated insert (see also polishing reaction above), 1 µl of restriction enzyme (*Srf* I, 5 U/ µl) and 1 µl of T4 DNA ligase (4 U/ µl). The reaction is incubated at room temperature for 1-2 hours and 2 µl removed and used to transform *E. coli.*

Protocol 3. PCR-based Site Directed Mutagenesis

Plasmid template DNA (approximately 0.5 pmole) is added to a PCR cocktail containing, in 25 µl of 1x mutagenesis buffer: (20 mM Tris HCl, pH 7.5; 8 mM $MgCl_2$; 40 µg/ml BSA); 12-20 pmole of each primer (one of which *must* contain a 5' phosphate), 250 µM each dNTP, 2.5 U *Taq* DNA polymerase, 2.5 U of *Taq* Extender (33, Stratagene). The PCR cycling parameters are 1 cycle of: 4 min at 94 °C, 2 min at 50 °C and 2 min at 72 °C; followed by 5-10 cycles of 1 min at 94 °C, 2 min at 54 °C and 1 min at 72 °C (*step 1*). The parental template DNA and the linear, mutagenesis-primer incorporating newly synthesized DNA are treated with *Dpn*I (10 U) and *Pfu* DNA polymerase (2.5U). This results in the *Dpn*I digestion of the *in vivo* methylated parental template and hybrid DNA (33) and the removal, by *Pfu* DNA polymerase, of the *Taq* DNA polymerase-extended base(s) on the linear PCR product. The reaction is incubated at 37 °C for 30 min and then transferred to 72 °C for an additional 30 min (*step 2*). Mutagenesis buffer (1x, 115 µl, containing 0.5 mM ATP) is added to the *Dpn*I-digested, *Pfu* DNA polymerase-polished PCR products. The solution is mixed and 10 µl is removed to a new microfuge tube and T4 DNA ligase (2-4 U) added. The ligation is incubated for greater than 60 min at 37 °C (*step 3*). The treated solution is transformed into competent *E. coli* (*step 4*).

Protocol 4. Colony PCR

Aliquot 50 µl of colony PCR reaction mix: 40 µl sterile, distilled water, 5 µl 10x PCR or ScreenTest buffer (35, 36; Stratagene), 0.4 µl 25 mM dNTP (6.25 mM *each* nucleoside triphosphate), 1-200 ng each PCR primer, 2.5 U *Taq* Extender (Stratagene, La Jolla, CA) and 2.5 U *Taq* DNA polymerase. Inoculate this reaction mix with a toothpick that has been stabbed into a colony from an agar plate. Overlay the inoculated mix with mineral oil and begin PCR. Initial cycling parameters may have to be adjusted for a particular primer set, but *general* conditions should include a 2-4 minute denaturation step at 94 °C followed by 2 minutes at 50 °C, and then 2 minutes at 72 °C. The first segment can then be followed by 25-30 more cycles using 1 minute at 94 °C followed by 2 minutes at 50 °C, and 1 minute at 72 °C. Note that these suggested times and temperatures may need to be adjusted for particular primer sets.

References

1. Hoof, T., Riordan, J.R. and Tummler, B. 1991. Quantitation of mRNA by the kinetic polymerase chain reaction assay: A tool for monitoring P-glycoprotein gene expression. Anal. Biochem. 196: 161-169.
2. Foulkes, N.S., Pandolfi de Rinaldis, P.P., Macdonnell, J., Cross, N.C.P. and Luzzatto, L. 1988. Polymerase chain reaction automated at low cost. Nucleic Acids Res. 16: 5687-5688.
3. Swerdlow, H., Dew-Jager K. and Gesteland, R. 1993. Rapid cycle sequencing in an air thermal cycler. BioTechniques. 15: 512-519.
4. Garner, H., Armstrong, B. and Lininger, D. 1993. High-throughput PCR. Biotechniques. 14: 112-115.

5. Gosden, J. and Hanratty, D. 1993. PCR *in situ*: A rapid alternative to *in situ* hybridization for mapping short, low copy number sequences without isotopes. BioTechniques. 15: 78-80.
6. Janoschek, R. and du Moulin, G.C. 1994. Ultraviolet disinfection in biotechnology: Myth vs. practice. BioPharm. Jan-Feb. 1994: 24-29.
7. Dycaico, M. and Mathur, S. 1991. Reduce PCR false positives using the Stratalinker UV crosslinker. Stratagies in Molec. Biol. 4: 39-40.
8. Longo, M., Berninger, M. and Hartley, J. 1990. Use of uracil DNA glycosylase to control carry-over contamination in polymerase chain reactions. Gene. 93: 125.
9. Saiki, R., Gelfand, D.H., Stoffel, S., Scharf, S.J., Higuchi, R., Horn, G.T., Mullis, K.B. and Erlich, H.A. 1988. Primer-directed enzymatic amplification of DNA with thermostable DNA polymerase. Science. 239: 487-491.
10. Lawyer, F.C., Stoffel, S., Saiki, R.K., Chang, S-Y., Landre, P.A., Abramson, R.D. and Gelfand, D.H. 1993. High-level expression, purification, and enzymatic characterization of full-length *Thermus aquaticus* DNA polymerase and a truncated form deficient in 5' to 3' exonuclease activity. PCR Meth. Applic. 2: 275-287.
11. Barnes, W.M. 1992. The fidelity of *Taq* polymerase catalyzing PCR is improved by an N-terminal deletion. Gene. 112: 29-35.
12. Mattila, P., Tonka, J., Tenkanen, T. and Pitkanen, K. 1991. Fidelity of DNA synthesis by the *Thermococcus litoralis* NDA polymerase, an extremely heat stable enzyme with proofreading activity. Nucleic Acids Res. 19: 4967-4976.
13. Kong, J., Kucera, R.B. and Jack, W.E. 1993. Characterization of a DNA polymerase from the hyperthermophile archaea *Thermococcus litoralis*. J. Biol. Chem. 268: 1965-1975.
14. Lundberg, K.S., Shoemaker, D.D., Adams, M.W.W., Short, J.M., Sorge, J.A. and Mathur, E.J. 1991. High-fidelity amplification using a thermostable DNA polymerase isolated from *Pyrococcus furiosus*. Gene. 108: 1-6.
15. Flaman, J.-M., Frebourg, T., Moreau, V., Charbonnier, F., Martin, C., Ishioka, C., Friend, S., and Iggo, R. 1994. A rapid PCR fidelity assay. Nucleic Acids Res. 22: 3259-3260.
16. Ruttimann, C., Cotoras, M., Zaldivar, J. and Vicuna, R. 1985. DNA polymerases from the extremely thermophilic bacterium *Thermus thermophilus* HB-8. Eur. J. Biochem. 149: 41-46.
17. Harrell, R. and Hart, R. 1994. Rapid preparation of *Thermus flavus* DNA polymerase. PCR Meth. Applic. 3: 372-375.
18. Kaledin, A.S., Slyusarenko, A.G. and Gorodetskii, S.I. 1981. Isolation and properties of DNA polymerase from the extremely thermophilic bacterium *Thermus flavus*. Biokhimiya. 46: 1576-1584.
19. Barnes, W.M. 1994. PCR amplification of up to 35-kb DNA with high fidelity and high yield from λ bacteriophage templates. Proc. Natl. Acad. Sci. USA. 91: 2216-2220.
20. Cheng, S., Fockler, C., Barnes, W.M. and Higuchi, R. 1994. Effective amplification of long targets from cloned inserts and human genomic DNA. Proc. Natl. Acad. Sci. USA. 91: 5695-5699.
21. Foord, O.S. and Rose, E.A. 1994. Long-distance PCR. PCR Meth. Applic. 3: S149-S161.
22. Hedden, V., Simcox, M., Scott, B., Cline, J., Nielson, K., Mathur, E. and K. Kretz. 1992. Superior sequencing: Cyclist Exo$^-$ *Pfu* DNA sequencing kit. Strategies in Molec. Biol. 5: 79.

23. Kretz, K., Callen, W. and V. Hedden. 1994. Cycle Sequencing. PCR Meth. Applic. 3: S107-112.
24. Costa, G.L., Grafsky, A. and Weiner, M.P. 1994. Cloning and Analysis of PCR-generated DNA fragments. PCR Meth. Applic. 3: 338-345.
25. Clark, J.M. 1988. Novel non-templated nucleotide addition reactions catalyzed by procaryotic and eucaryotic DNA polymerases. Nucleic Acids Res. 16: 9677-9686.
26. Hu, G. 1993. DNA polymerase-catalyzed addition of nontemplated extra nucleotides to the 3'end of a DNA fragment. DNA Cell Biol. 12: 763-770.
27. Liu, Z. and Schwartz, L. 1992. An efficient method for blunt-end ligation of PCR product. BioTechniques. 12: 28-30.
28. Bauer, J., Deely, D Braman, J., Viola, J. and Weiner, M.P. 1992. pCR-Script SK(+) cloning system: a simple and fast method for PCR cloning. Strategies in Molec. Biol. 5: 62-65.
29. Costa, G.L., and Weiner, M.P. 1994. Polishing with T4 or *Pfu* polymerase increases the efficiency of cloning PCR fragments. Nucleic Acids Res. 22: 2423.
30. Costa, G.L., and Weiner, M.P. 1994. Increased cloning efficiency with the PCR polishing kit. Strategies in Molec. Biol. 7: 47-48.
31. Simcox, T., Marsh, S., Gross, E., Lernhardt, L., Davis, S. and Simcox, M. 1991. *Srf* I, a new type-II restriction endnuclease that recognizes the octanucleotide sequence, 5'-GCCCGGGC-3'. Gene. 109: 121-123.
32. Weiner, M.P., Costa, G.L., Schoettlin, W., Cline, J., Mathur, E. and Bauer, J. 1994. Site-directed mutagenesis of double-stranded DNA by the polymerase chain reaction. Gene. 151: 119-123.
33. Nelson, M. and McClelland, M. 1992. The use of DNA methyltransferase/endonuclease enzyme combinations for megabase mapping of chromosomes. Meth. Enzymol. 216: 279-303.
34. Nielson, K.B., Schoettlin, W., Bauer, J.C. and Mathur, E. 1994. *Taq* ExtenderPCR additive for improved length, yield and reliability of PCR products. Strategies in Molec. Biol. 7: 27.
35. Costa, G.L., and Weiner, M.P. 1994. ScreenTest recombinant screening in one day. Strategies in Molec. Biol. 7: 35-37.
36. Costa, G.L. and Weiner, M.P. 1994. ScreenTest colony-PCR screening: questions and answers. Stratagies in Molec. Biol. 7: 72-74.

From: *Molecular Biology: Current Innovations and Future Trends.*
ISBN 1-898486-01-8 ©1995 Horizon Scientific Press, Wymondham, U.K.

2

THERMAL CYCLE SEQUENCING

Keith A. Kretz

Abstract

The acquisition of DNA sequence information is an integral part of many research projects. The goals of the Human Genome Project alone include determining the sequence of the 3 billion base pairs of DNA in the human genome as well as several model organisms (1, 2). To that end a number of improvements have been made in sequencing strategies and data acquisition which allow sequence data to be obtained more efficiently. One of these improvements was the development of thermal cycle sequencing, or linear amplification sequencing, as it is sometimes known. This method utilizes a thermostable DNA polymerase in a temperature cycling format to perform multiple rounds of dideoxynucleotide sequencing on the template. The advantages include stronger signals from smaller amounts of template and the ability to sequence previously unreliable templates.

Introduction

There are many methods for sequencing DNA. Those which were the most widely used are the chemical cleavage method of Maxam and Gilbert (3) and the dideoxynucleotide terminator chemistry of Sanger *et al.* (4). The Sanger approach became the preferred method in many laboratories when Tabor and Richardson developed modified T7 DNA polymerase (5). While using modified T7 DNA polymerase provides high quality sequence data, there are some disadvantages associated with this approach, namely the need for a large amount of high quality DNA template and the thermolabile nature of the enzyme. The first report of the use of *Taq* DNA polymerase, an enzyme which could withstand repeated heating to 94-95 °C, in a dideoxynucleotide terminator format opened the door for the development of cycle sequencing. Cycle sequencing is the combination of temperature cycling with dideoxynucleotide terminator sequencing (6, 7). While cycle sequencing does not generate a logarithmic increase in yield, the linear amplification obtained in 20-30 rounds of temperature cycling significantly increases the signal and allows the use of much smaller amounts of DNA template. Repeated rounds of denaturation and primer annealing also allow a much wider range of DNA (cosmids, lambda DNA and PCR products in addition to the usual M13 and plasmid DNA) to be used as sequencing templates.

Dideoxynucleotide Terminator Sequencing

Sanger *et al.* (4) demonstrated that the Klenow fragment of *Escherichia coli* DNA polymerase I could incorporate 2',3'-dideoxynucleoside triphosphates into a growing strand of DNA resulting in a base specific termination event which could be used to sequence DNA. The method is simple and robust. A primer is annealed specifically to a template in the presence of DNA polymerase and a mixture of all four of the deoxynucleotides and one of the dideoxynucleotides. When present in an appropriate ratio, dideoxynucleoside triphosphate incorporation results in a series of specifically terminated fragments. This is repeated with each of the four dideoxynucleoside triphosphates. The newly synthesized fragments are then separated by denaturing polyacrylamide gel electrophoresis and the sequence determined from the resulting ladder.

The Klenow fragment of *E. coli* DNA polymerase I was the enzyme of choice for many years until the description of a modified T7 DNA polymerase by Tabor and Richardson (5). It was discovered that 3'-5' exonuclease deficient T7 DNA polymerase had several properties which made it a superior enzyme for DNA sequencing with chain terminators. The modified polymerase was highly processive and efficiently incorporated the dideoxynucleotide chain terminators, resulting in uniform band intensities over a wide range of synthesis sizes. Further work by Tabor and Richardson described improvements in the quality of DNA sequence with the use of different metal ions, specifically manganese (8), and the inclusion of inorganic pyrophosphatase (9). The inclusion of manganese in the reaction buffer reduces the discrimination of the dideoxynucleotides by the polymerase resulting in highly uniform band intensities (8). The addition of pyrophosphatase prevents pyrophosphorolysis by bacteriophage T7 DNA polymerase which may lead to the degradation of specific dideoxynucleotide terminated fragments visualized on DNA sequencing gels (9).

DNA sequencing reaction products were initially detected using radioactively labeled primers or by the incorporation of an alpha-labeled dNTP. Traditionally, ^{32}P was the isotope used in DNA sequencing until the introduction of ^{35}S. ^{35}S has a longer half-life and has a lower energy β-emission than ^{32}P resulting in sharper bands on the autoradiographs and decreased risk to laboratory personnel. In 1991, Zagursky *et al.* (10, 11) reported the use of ^{33}P labeled nucleotides in DNA sequencing. ^{33}P has physical characteristics intermediate between those of ^{32}P and ^{35}S. The intermediate energy of the β-emission results in sharp bands like those of ^{35}S but more intense, making it an excellent alternative for sequencing. In recent years a number of non-radioactive detection strategies have been developed. For example, biotinylated primers may be detected by interaction with streptavidin-horseradish peroxidase or streptavidin-alkaline phosphatase complexes followed by colorimetric or chemiluminescent detection. These detection strategies are particularly powerful when coupled to a multiplex DNA sequencing strategy.

Multiplex DNA sequencing utilizes a pooling scheme early in the sequencing strategy to reduce the amount of effort spent in template preparation, sequencing and gel electrophoresis (12). A set of "multiplex" vectors are used to clone the fragments which need to be sequenced. Each multiplex vector differs from the others in the "tag" region on either side of the cloning site. Pools of clones are prepared, one colony from each vector library. Each pool of clones is grown overnight, DNA is prepared and sequenced. After separating the sequencing products by electrophoresis, the reaction products are transferred to a membrane. The sequencing ladders are detected by hybridization of primers to the unique tag sequences present in the cloning vectors. Radioactively labeled oligonucleotides as well as hapten-labeled oligos or enzyme-conjugated oligos can be used for detection. The probe can then be stripped from the membrane and additional sets of sequencing reactions can be detected with repeated rounds of hybridization with different labeled primers. This process is repeated for each of the tag sequences present in the pool. Coupling the multiplex DNA sequencing technique to direct blotting electrophoresis makes this technique very useful (13). Direct blotting electrophoresis utilizes a polyacrylamide gel for separation of the sequencing reaction products which are directly deposited onto a membrane which is moved across the bottom of the gel (14).

A number of automated DNA sequencing systems have been developed (15-19). All are based on the labeling of the sequencing reaction products with fluorophores which are excited by a laser near the bottom of the gel. The emitted light is collected and analyzed by a computer which determines the base sequence. The systems are of two main types. One color, four lane machines use a single species of fluorophore, such as fluorescein, and run the four termination reactions in adjacent lanes in the gel. The advantage of this approach is that only one fluorophore needs to be excited and detected, however four separate sequencing reactions need to be run and each requires a lane on the gel. Generally a labeled primer is used but Voss *et al.* (20, 21) have reported the use of fluorescein-dATP as an internal label. Four color, one lane sequencing has the advantage that all four sets of termination reactions can be run in a single lane on the sequencing gel because each is labeled with a different fluorophore. Several criteria need to be met to make the fluorophores useful in this format (15): (i) The emission spectra have to be well resolved from each other. (ii) The dyes must be highly fluorescent to achieve the necessary sensitivity. (iii) The dyes can not significantly distort the electrophoretic mobility of the labeled DNA fragments. Computer software has been created which compensates for these differences in mobility when determining the base sequence. The four color, one lane sequencing reaction chemistries have been designed in two different ways. Either dye-primer (15, 16), where the dyes are attached to the primers, or dye-terminator (18) chemistries can be used. Dye-terminators are dideoxynucleotide analogs containing the fluorophores. In addition to being run in a single lane on the gel, the dye-terminator labeling approach offers several advantages. Generation of all four sets of DNA sequencing fragments can be carried out simultaneously in a single reaction since only the terminating nucleotide carries the tag. In addition, many polymerase pausing artifacts are eliminated since only those fragments resulting from true termination events carry a fluorescent tag and are detected by the instrument (18).

Cycle Sequencing

The ability of *Taq* DNA polymerase to withstand the temperatures necessary to denature DNA resulted in the wide spread utilization of the polymerase chain reaction (PCR; 22). The relatively high temperature optimum of this polymerase was quickly exploited for use in chain termination sequencing at higher temperatures, 50-60 °C, in an attempt to disrupt template secondary structures (23). Soon, chain-termination sequencing with *Taq* was linked to temperature cycling to take advantage of the benefits of

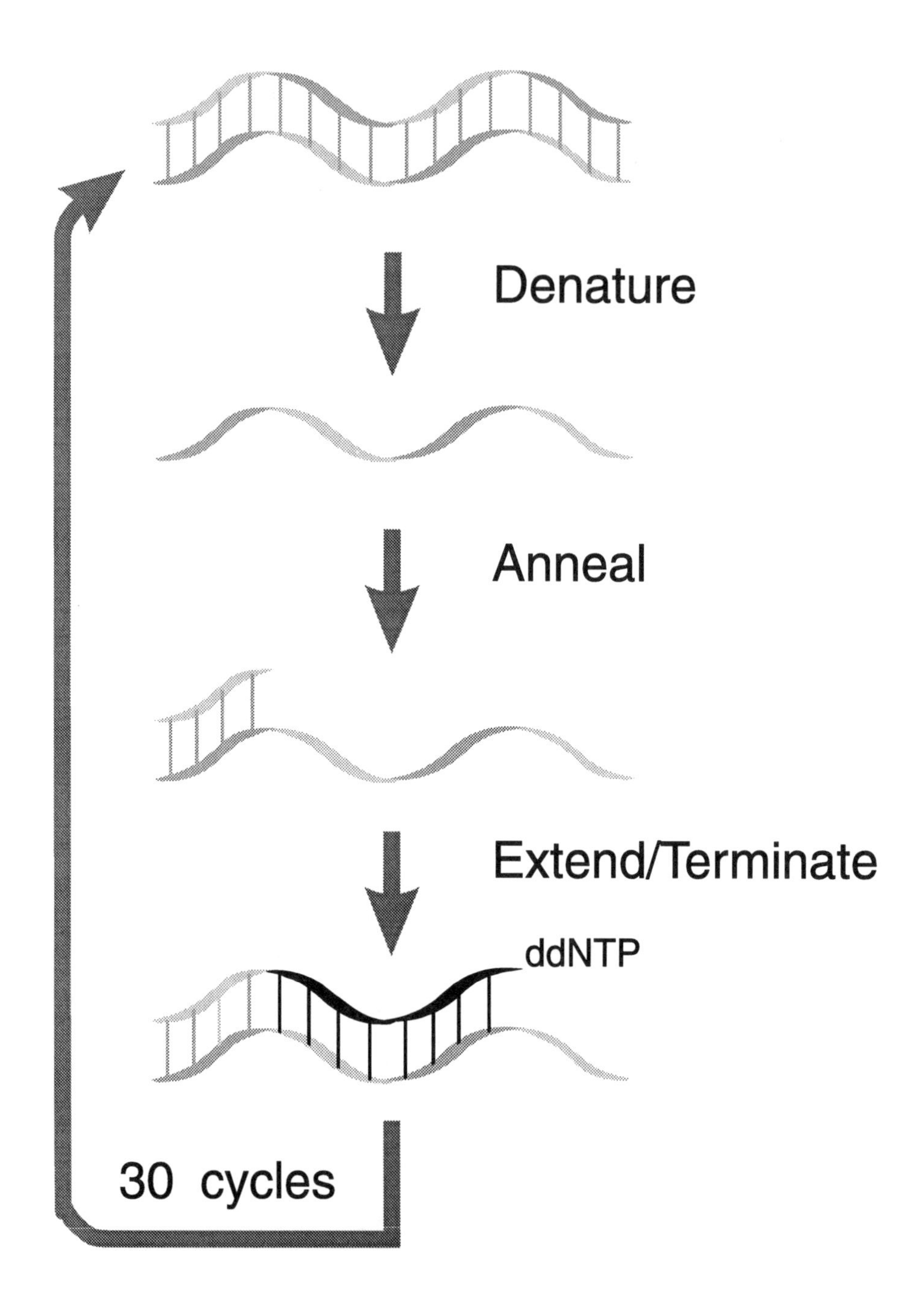

Figure 1. Cycle Sequencing. Cycle sequencing consists of consecutive steps of high temperature denaturation, annealing at a temperature optimal for the primer/template set and extension at a temperature optimal for the DNA polymerase used. This process can be repeated 30 or more times to increase the amount of product to be run on a sequencing gel.

each (6, 7). This hybrid technique was named cycle sequencing, or linear amplification sequencing. Cycle sequencing utilizes a temperature cycling machine to perform repeated rounds of denaturation, primer annealing and primer extension analogous to PCR (Figure 1). The differences between PCR and cycle sequencing include the following: (i) only one primer is present in cycle sequencing. (ii) four separate reactions are run, one for each of the four dideoxynucleotide terminations unless dye-terminator detection chemistry is used (see below). Temperature cycling results in the linear amplification of the sequencing products resulting in greater signal strength. Cycling the sequencing reactions results in several advantages: (i) considerably less DNA template is required to obtain sufficient signal, (ii) because smaller amounts of template are added, fewer impurities are introduced, meaning less template purification is required for manual cycle sequencing, (iii) the high temperature at which the sequencing reactions are run and the multiple heat-denaturation steps, allow double stranded templates such as plasmids, cosmids, lambda DNA and PCR products to be sequenced reliably without a separate denaturation step. Many commercial kits are available to perform cycle sequencing reactions and an extensive list is compiled in reference 24.

Sequencing Templates

Cycle sequencing enables researchers to sequence much smaller amounts of DNA with little or no purification if manual protocols are followed. However, it should be noted that the purity of the template may affect the quality of the sequence data. Protocols and detection chemistries used in automated DNA sequencing instruments require the highest quality templates. Previously inconsistent DNA templates such as PCR products, cosmids, and lambda DNA can now be sequenced routinely using the cycle sequencing technique. Plasmids, M13, cosmids and lambda DNA purified by any standard small- or large-scale protocol are suitable templates (25).

When a PCR product is to be used as the sequencing template, a number of very important points need to be considered (26-29). The most important decision is whether to sequence the PCR reaction product directly or after cloning into a vector. It is well documented that the error rate of *Taq* DNA polymerase is such that many of the reaction products may contain misincorporation errors (26). Therefore, if the reaction products are cloned into a vector prior to sequencing, multiple clones must be sequenced to determine a consensus sequence. If the PCR reaction product is sequenced directly, any sequence differences in individual products will be masked by the consensus. In the worst case, a single template molecule is present and there is a misincorporation in the first round of PCR. This will give rise to that particular misincorporation being present in 25% of the final product molecules. Generally, this would not be a problem when reading the sequencing gel. In most cases, there will be many template molecules present at the start of the PCR reaction and any individual mutation will represent a very small fraction of the final product.

While it is apparent that in most cases it is best to sequence the PCR reaction product directly, this is technically more difficult than sequencing DNA from plasmid or M13 clones. This is due to the efficient reannealing of the linear, double-stranded DNA molecules after denaturation. The PCR reaction also contains primers and excess nucleotides which may interfere with subsequent sequencing reactions. Many methods have been devised to address these problems such as: (i) ^{32}P end-labeled internal primers

(30), (ii) ^{32}P end-labeled primers used after ultrafiltration or column removal of the amplification primers and nucleotides (31), (iii) asymmetric PCR amplification to produce excess single-stranded DNA (32), (iv) production of ssDNA after PCR amplification using exonuclease digestion (33), (v) PCR amplification with biotinylated primers to allow the capture of ssDNA for sequencing (34), (vi) genomic amplification with transcript sequencing (35), (vii) the use of a blocking primer (36).

While these methods may be useful, they are time-consuming and generally unnecessary if a cycle sequencing protocol is used. The repeated thermal denaturation and annealing steps of a cycle sequencing reaction provide ample opportunity for primer annealing and extension. Because the presence of unused primers and nucleotides in the PCR reaction product may disrupt the carefully optimized sequencing reaction conditions, the lowest concentration of nucleotides and primers which still result in efficient amplification should be used. In addition, the amplification afforded by PCR is such that only a small fraction of the final reaction product (0.1-2 μl) is generally necessary as the template in a cycle sequencing reaction. Carryover under these conditions is not generally deleterious to the sequencing reaction.

The other significant problem encountered in direct sequencing of PCR products is the presence of non-specific amplification products. If one of the PCR primers is to be used as the sequencing primer, it will bind with equal affinity to all of the PCR reaction products (specific or non-specific) which have incorporated it. There are several ways to overcome this problem. The easiest solution is to modify the PCR reaction parameters such that only the specific product is synthesized. If this is not possible, the reaction products must be separated on an agarose gel and the band of interest recovered. These products may be recovered from the gel by a number of methods including silica-based resin extraction, enzymatic digestion of the agarose (37), or a modified freeze-squeeze protocol (38). Alternatively, the PCR product can be isolated in low-melting temperature agarose. After melting the gel slice, a portion of the band can be added directly to the sequencing mixture (39). To circumvent the need for gel isolation, a third primer which anneals to the PCR product internal to the amplification primers can be used to sequence the PCR product in an end-labeled format (30). This primer will only anneal to the specific PCR product and can therefore give high quality sequence data from a mixed template population.

Because of the stringency of primer annealing and increased signal strength afforded by cycle sequencing, larger sequencing templates can now be sequenced directly. In an attempt to determine the upper limit for size of the sequencing template, unfractionated, high molecular weight *E. coli* genomic DNA has been used as the template in a cycle sequencing reaction. Useful sequence information for 100-150 bases was obtained from this reaction (40). Additional experiments showed that this is not yet a reliable technique but it does indicate that further improvements in cycle sequencing will result in the ability to sequence much larger templates. This would greatly reduce the amount of effort required to complete large scale sequencing projects.

Conclusions and Future Trends

The dideoxynucleotide sequencing technique has evolved into a versatile, efficient and robust method. The discovery of new DNA polymerases, associated enzymes and modified nucleotides has rapidly expanded the limits of this powerful sequencing technique. The use of thermostable enzymes in cycle sequencing formats and the use of inorganic pyrophosphatase have increased the quality of the sequence data. Automated sequencing systems utilizing fluorescent labels are safer and provide increased capacity. Similar improvements in the upstream processing of DNA templates include robotic workstations which can isolate high quality template DNA and prepare the sequencing reactions. Gel electrophoresis has been improved with the introduction of ultrathin, temperature-controlled gels which run at very high speed and provide increased resolution. Direct blotting electrophoresis is an advantage for those who choose to use the multiplex sequencing methods. Current sequence analysis techniques and the DNA databases are stretched to the limit trying to keep up with the ever increasing amount of data provided by these improving techniques.

The future is even more promising with the possibility of incremental improvements in existing techniques and several revolutionary new approaches. Capillary gel electrophoresis using replaceable linear (non-crosslinked) polyacrylamide is receiving much attention because of the ease of use, capacity and speed of separation (41). Directed sequencing approaches like primer-walking (42, 43) and transposon-facilitated sequencing (44, 45) are improving to the point that they are viable alternatives to shotgun sequencing approaches. Another recent technology is sequencing by hybridization which holds great promise for the analysis of sequence variation (46-48). Some researchers are attempting to use mass spectrometry or scanning tunneling microscopy to determine the sequence of DNA while other research efforts are aimed at "single-molecule" detection strategies (49).

Great improvements have been made in the methods used to obtain DNA sequence information while several revolutionary new methods offer the hope of dramatic increases in the rate at which this information can be obtained. An ever increasing amount of DNA sequence information is available to aid the study of more complex biological systems and to be used in clinical diagnostic laboratories. This information will be invaluable in the future.

Protocols

Typical Protocol for Manual Cycle Sequencing (Radioactive Labeling and *Taq* DNA Polymerase)

This protocol and these solutions are sold commercially (Cyclist™ *Taq* DNA Sequencing Kit, Stratagene Cloning Systems, La Jolla, CA).

1. Add 3 µl of the appropriate termination mix to each of 4 tubes. Cap the tubes and keep on ice.

2. For each DNA template, combine the following reaction components on ice and mix gently but thoroughly by pipeting:
   - template
   - primer (2-5 pmole; 10-25 ng of a 17mer)
   - 4 µl 10X sequencing buffer
   - 10 µCi [α-^{32}P]-dATP or [α-^{33}P]-dATP (NEN-DuPont, Boston, MA; not necessary if labeled primers are used)
   - 2-5 U *Taq* DNA polymerase
   - water to a final volume of 30 µl
3. Aliquot 7 µl of the reaction mixture from step 2 into each of the 4 tubes containing termination mix from step 1.
4. Overlay the reactions with 15 µl of silicone oil (Sigma Chemical Company, St. Louis, MO) or mineral oil (Sigma Chemical Company, St. Louis, MO), if necessary.
5. Optimum cycling parameters will vary depending on template, primer and temperature cycler used. A useful starting profile is given below:
   30 cycles each of
   95 °C for 30 seconds
   50 °C for 30 seconds
   72 °C for 60 seconds
6. Add 5 µl of stop dye below the oil overlay and mix by pipeting.
7. Heat denature the samples at ≥80 °C for 2-5 minutes immediately prior to loading 1-3 µl onto a sequencing gel e.g. 6% polyacrylamide, 7 M urea, 1X TBE (Stratagene Cloning Systems, La Jolla, CA; 25).

Solutions

10X Sequencing Buffer:
- 100 mM Tris-HCl (pH 8.8), 500 mM KCl, 40 mM $MgCl_2$, 0.01% gelatin, 20 µM dATP
- 50 µM dCTP, 50 µM dGTP, 50 µM dTTP.

Termination mixes:
- ddATP (600 µM)
- ddCTP (600 µM)
- ddGTP (100 µM)
- ddTTP (1000 µM)

Stop dye:
- 95% formamide, 20 mM EDTA, 0.05% bromophenol blue, 0.05% xylene cyanol.

Note that these solutions were designed to allow the user to manipulate the dNTP/ddNTP ratio. When using the recommended amounts of all components the reaction products extend into the 400-500 nt range. By using less ddNTP termination mix (1µl) in each reaction, these products can be extended to over 1000 nt. Conversely, if you are sequencing a short PCR fragment, using more ddNTP termination mix (5-6 µl) in the reaction will result in quicker termination and more sequence information close to the primer.

Template Preparation for Manual Cycle Sequencing

Plasmids, M13, cosmids and lambda DNA purified by any standard small- or large-scale protocol are suitable templates (25). Typically 50-200 fmoles of plasmid (100-400 ng of a 3kb plasmid) or cosmid DNA are used in the sequencing reactions, however 1-500 fmoles of plasmid template have been used successfully. For M13 and lambda DNA, 10-100 fmoles should be used as template.

As described above, PCR products may be sequenced directly if a single product is present. Generally 0.1-2 µl of the reaction product is sufficient template for the cycle sequencing reaction. Note that it is very easy to overload the sequencing reaction with a short PCR product. Whenever possible, quantitate the DNA and use 10-100 fmoles in the reaction to reduce background problems. If the sequence obtained using unpurified PCR template has an unacceptably high background, unused amplification primers and nucleotides may be removed by gel purification, purification on a silica resin, or by selective precipitation of the PCR product as follows:

1. Mix an aliquot of the PCR reaction with an equal volume of 5 M ammonium acetate.
2. Add an equal volume (PCR reaction plus ammonium acetate) of room temperature isopropyl alcohol.
3. Incubate at room temperature for 10 minutes.
4. Microcentrifuge for 15 minutes.
5. Remove the isopropyl alcohol supernatant and resuspend the pellet (usually invisible) in water or TE to a volume equal to the original PCR volume.
6. Use an appropriate amount as template.

If gel isolation is required due to the presence of non-specific amplification products, the band of interest may be recovered as described above (37-39).

Methods have also been developed to sequence plasmid DNA directly from colonies and phage DNA from plaques (50). High copy number plasmids may be sequenced directly from colonies using the following protocol. This protocol requires the use of ^{32}P and overnight exposure or a 2-3 day exposure with ^{33}P.

1. Pick a colony from an agar plate and boil it in 25 µl of TE (5 mM Tris-HCl, pH 8.0, 0.1 mM EDTA) for 5 minutes.
2. Vortex well and cool on ice. Remove the cell debris by centrifugation in a microcentrifuge for 1 minute.
3. Use 10 µl of the supernatant as the template in subsequent sequencing reactions.

If the plasmid is present in low copy number, a 1 ml overnight liquid culture may be prepared. Collect 50 to 100 µl of the cells by centrifugation for 1 minute and resuspend in 25 µl of TE. Boil the resuspended cells and prepare as described above. While the use of plasmid DNA obtained by boiling cells in this manner will not be of the highest quality (150-200 bases of readable sequence), the technique is particularly useful when screening large numbers of colonies as may be the case with site-specific mutagenesis or cloning projects.

The DNA present in M13 or lambda plaques may be sequenced directly in a manner analogous to that for plasmids from colonies. Again, this protocol requires the use of ^{32}P and overnight exposure or a 2-3 day exposure with ^{33}P.

1. Cut closely around the plaque of interest with a scalpel and peel off *just the top agarose layer.*
2. Boil the top agarose containing the plaque in 25 µl of TE for 5 minutes.
3. Vortex well and cool on ice. Remove any remaining pieces of agarose by centrifugation for 1 minute in a microcentrifuge.
4. Use 10 µl of the supernatant as the template in subsequent sequencing reactions.

References

1. Watson, J.D. 1990. The Human Genome Project: Past, present, and future. Science. 248: 44-49.
2. Cantor, C.R. 1990. Orchestrating the Human Genome Project. Science. 248: 49-51.
3. Maxam, A.M. and Gilbert, W. 1977. A new method for sequencing DNA. Proc. Natl. Acad. Sci. USA. 74: 560-564.
4. Sanger, F., Nicklen, S. and Coulson, A.R. 1977. DNA sequencing with chain-terminating inhibitors. Proc. Natl. Acad. Sci. USA. 74: 5463-5467.
5. Tabor, S. and Richardson, C.C. 1987. DNA sequence analysis with a modified bacteriophage T7 DNA polymerase. Proc. Natl. Acad. Sci. USA. 84: 4767-4771.
6. Murray, V. 1989. Improved double-stranded DNA sequencing using the linear polymerase chain reaction. Nucleic Acids Res. 17: 8889.
7. Carothers, A.M., Urlaub, G., Mucha, J., Grunberger, D. and Chasin, L.A. 1989. Point mutation analysis in a mammalian gene: Rapid preparation of total RNA, PCR amplification of cDNA, and *Taq* sequencing by a novel method. BioTechniques. 7: 494-499.
8. Tabor, S. and Richardson, C.C. 1989. Effect of manganese ions on the incorporation of dideoxynucleotides by bacteriophage T7 DNA polymerase and *Escherichia coli* DNA polymerase I. Proc. Natl. Acad. Sci. USA. 86: 4076-4080.
9. Tabor, S. and Richardson, C.C. 1990. DNA sequence analysis with a modified bacteriophage T7 DNA polymerase. J. Biol. Chem. 265: 8322-8328.
10. Zagursky, R.J., Conway, P.S. and Kashdan, M.A. 1991. Use of ^{33}P for Sanger DNA sequencing. BioTechniques. 11: 36-38.
11. Evans, M.R. and Read, C.A. 1992. ^{32}P, ^{33}P and ^{35}S: Selecting a label for nucleic acid analysis. Nature. 358: 520-521.
12. Church, G.M. and Kieffer-Higgins, S. 1988. Multiplex DNA sequencing. Science. 240: 185-188.
13. Richterich, P. and Church, G.M. 1993. DNA sequencing with direct transfer electrophoresis and nonradioactive detection. Methods in Enzymol. 218:187-222.
14. Beck, S. and Pohl, F.M. 1984. DNA sequencing with direct blotting electrophoresis. EMBO J. 3: 2905-2909.
15. Smith, L.M., Sanders, J.Z., Kaiser, R.J., Hughes, P., Dodd, C., Connell, C.R., Heiner, C., Kent, S.B.H. and Hood, L.E. 1986. Fluorescence detection in automated DNA sequence analysis. Nature. 321: 674-679.

16. Kaiser, R., Hunkapiller, T., Heiner, C. and Hood, L. 1993. Specific primer-directed DNA sequence analysis using automated fluorescence detection and labeled primers. Meth. Enzymol. 218: 122-153.
17. Ansorge, W., Sproat, B.S., Stegemann, J. and Schwager, C. 1986. A non-radioactive automated method for DNA sequence determination. J. Biochem. Biophys. Methods. 13: 315-323.
18. Prober, J.M., Trainor, G.L., Dam, R.J., Hobbs, F.W., Robertson, C.W., Zagursky, R.J., Cocuzza, A.J., Jensen, M.A. and Baumeister, K. 1987. A system for rapid DNA sequencing with fluorescent chain-terminating dideoxynucleotides. Science. 238: 336-341.
19. Middendorf, L.R., Bruce, J.C., Bruce, R.C., Eckles R.D., Grone, D.L., Roemer, S.C., Sloniker, G.D., Steffens, D.L., Sutter, S.L., Brumbaugh, J.A. and Patonay, G. 1992. Continuous, on-line DNA sequencing using a versatile infrared laser scanner/electrophoresis apparatus. Electrophoresis. 13: 487-494.
20. Voss, H., Wiemann, S., Wirkner, U., Schwager, C., Zimmermann, J., Stegemann, J., Erfle, H., Hewitt, N.A., Rupp, T. and Ansorge, W. 1992. Automated DNA sequencing system resolving 1,000 bases with fluorescein-15-dATP as internal label. Meth .Mol. Cell. Biol. 3: 153-155.
21. Voss, H., Wiemann, S., Grothues, D., Sensen, C., Zimmermann, J., Schwager, C., Stegemann, J., Erfle, H., Rupp, T. and Ansorge, W. 1993. Low redundancy sequencing: Primer walking on plasmid and cosmid DNA using fluorescein-dATP as internal label. BioTechniques. 15: 714-721.
22. Saiki, R.K., Gelfand, D.H., Stoffel, S., Scharf, S.J., Higuchi, R., Horn, G.T. and Erlich, H.A. 1988. Primer-directed enzymatic amplification of DNA with thermostable DNA polymerase. Science. 239: 487-491.
23. Innis, M.A., Myambo, K.B., Gelfand, D.H. and Brow, M.A.D. 1988. DNA sequencing with *Thermus aquaticus* DNA polymerase and direct sequencing of polymerase chain reaction-amplified DNA. Proc. Natl. Acad. Sci. USA. 85: 9436-9440.
24. Roberts, S.S. Thermostable DNA polymerases heat up DNA sequencing. 1992. J. NIH Res. 4: 89-94.
25. Sambrook, J., Fritsch, E.F. and Maniatis, T. 1989. Molecular Cloning: A Laboratory Manual. Cold Spring Harbor Laboratory Press, Cold Spring Harbor, New York.
26. Andersson, B. and Gibbs, R.A. 1994. PCR and DNA sequencing. In PCR: The Polymerase Chain Reaction. K.B. Mullis, F. Ferre and R.A. Gibbs, eds. Birkhauser, Boston. p. 201-213.
27. Bevan, I.S., Rapley, R. and Walker, M.R. 1992. Sequencing of PCR-amplified DNA. PCR Meth. Applic. 1: 222-228.
28. Rao, V.B. 1994. Direct sequencing of polymerase chain reaction-amplified DNA. Anal. Biochem. 216: 1-14.
29. Thomas, W.K. and Kocher, T.D. 1993. Sequencing of polymerase chain reaction-amplified DNAs. Meth. Enzymol. 224: 391-399.
30. Wrischnik, L.A., Higuchi, R.G., Stoneking, M., Erlich, H.A., Arnheim, N., and Wilson, A.C. 1987. Length mutations in human mitochondrial DNA: Direct sequencing of enzymatically amplified DNA. Nucleic Acids Res. 15: 529-542.
31. Wong, C., Dowling, C.E., Saiki, R.K., Higuchi, R.G., Erlich, H.A. and Kazazian, H.H. 1987. Characterization of β-thalassaemia mutations using direct genomic sequencing of amplified single copy DNA. Nature. 330: 384-386.

32. Gyllensten, U.B. and Erlich, H.A. 1988. Generation of single-stranded DNA by the polymerase chain reaction and its application to direct sequencing of the *HLA-DQA* locus. Proc. Natl. Acad. Sci. USA. 85: 7652-7656.
33. Higuchi, R.G. and Ochman, H. 1989. Production of single-stranded DNA templates by exonuclease digestion following the polymerase chain reaction. Nucleic Acids Res. 17: 5865.
34. Mitchell, L.G. and Merril, C.R. 1989. Affinity generation of single-stranded DNA for dideoxy sequencing following the polymerase chain reaction. Anal. Biochem. 178: 239-242.
35. Stoflet, E.S., Koeberl, D.D., Sarkar, G. and Sommer, S.S. 1988. Genomic amplification with transcript sequencing. Science. 239: 491-494.
36. Gyllensten, U.B. 1989. PCR and DNA sequencing. BioTechniques. 7: 700-708.
37. Yaphe, W. 1957. The use of agarase from *Pseudomonas atlantica* and the identification of agar in marine algae (*Rhodophyceae*). Can. J. Microbiol. 3: 987-993.
38. Hedden, V., Callen, W. and Kretz, K. 1992. Freeze-N-Spin™ Filter Cups for Fast DNA Recovery. Strategies Molec. Biol. 5: 80.
39. Kretz, K.A. and O'Brien, J.S. 1993. Direct sequencing of PCR products from low-melting temperature agarose. Meth. Enzymol. 218: 72-79.
40. Kretz, K.A., Callen, W. and Hedden, V. 1994. Cycle sequencing. PCR Meth. Applic. 3: S107-S112.
41. Ruiz-Martinez, M.C., Berka, J, Belenkii, A., Foret, F., Miller, A.W. and Karger, B.L. 1993. DNA sequencing by capillary electrophoresis with replaceable linear polyacrylamide and laser-induced fluorescence detection. Anal. Chem. 65: 2851-2858.
42. Kieleczawa, J., Dunn, J.J. and Studier, F.W. 1992. DNA sequencing by primer walking with strings of contiguous hexamers. Science. 258: 1787-1791.
43. Kotler, L.E., Zevin-Sonkin, D., Sobolev, I.A., Beskin, A.D. and Ulanovsky, L.E. 1993. DNA sequencing: Modular primers assembled from a library of hexamers or pentamers. Proc. Natl. Acad. Sci. USA. 90: 4241-4245.
44. Strathmann, M., Hamilton, B.A., Mayeda, C.A., Simon, M.I., Meyerowitz, E.M. and Palazzolo, M.J. 1991. Transposon-facilitated DNA sequencing. Proc. Natl. Acad. Sci. USA. 88: 1247-1250.
45. Kasai, H., Isono, S., Kitakawa, M., Mineno, J., Akiyama, H., Kurnit, D.M., Berg, D.E. and Isono, K. 1992. Efficient large-scale sequencing of the *Escherichia coli* genome: Implementation of a transposon- and PCR-based strategy for the analysis of ordered λ phage clones. Nucleic Acids Res. 20: 6509-6515.
46. Drmanac, R., Labat, I., Brukner, I. and Crkvenjakov, R. 1989. Sequencing of megabase plus DNA by hybridization: Theory of the method. Genomics. 4: 114-128.
47. Maskos, U. and Southern, E. 1993. A novel method for the parallel analysis of multiple mutations in multiple samples. Nucleic Acids Res. 21: 2269-2270.
48. Pease, A.C., Solas, D., Sullivan, E.J., Cronin, M.T., Holmes, C.P. and Fodor, S.P.A. 1994. Light-generated oligonucleotide arrays for rapid DNA sequence analysis. Proc. Natl. Acad. Sci. USA. 91: 5022-5026.

49. Davis, L.M., Fairfield, E.R., Harger, C.A., Jett, J.H., Hahn, J.H., Keller, R.A., Krakowski, L.A., Marrone, B.L., Martin, J.C., Nutter, H.L., Ratliff, R.L., Shera, E.B., Simpson, D.J. and Soper, S.A. 1991. Rapid DNA sequencing based upon single molecule detection. Genet. Anal. Tech. Appl. 8: 1-7.
50. Krishnan, B.R., Blakesley, R.W. and Berg, D.E. 1991. Linear amplification DNA sequencing directly from single phage plaques and bacterial colonies. Nucleic Acids Res. 19: 1153.

From: *Molecular Biology: Current Innovations and Future Trends.*
ISBN 1-898486-01-8 ©1995 Horizon Scientific Press, Wymondham, U.K.

3

MINI-PREP PLASMID DNA ISOLATION AND PURIFICATION USING SILICA-BASED RESINS

Paul N. Hengen

Abstract

It has been common practice in the past to use large quantities of starting material for experimental molecular biology techniques; however, rapid advancements in subcloning and sequencing procedures on a mini-scale have shown that many of these methods no longer require a substantial amount of purified DNA. Due to the high quality of many commercially available purified enzymes and reagents, and the development of many new mini-isolation techniques and prepackaged kits, high quality mini-prep plasmid DNA can now be had for far less investment of time, labour, and materials. A significant reduction in the time researchers spend on purifying their DNA samples has led to an increase in productivity for experimental biologists, one testimony to this being the rapid appearance and availability of numerous new cloning vectors. In this chapter, the different techniques for the isolation and purification of mini-prep plasmid DNA are reviewed and attention is focused on a rapid and inexpensive method utilizing the binding of DNA to diatomaceous earth and other silicates.

Introduction

Isolation of Plasmid DNA

The standard cesium chloride - ethidium bromide centrifugation technique (1), has now become the old grey mare of plasmid DNA isolation and purification. Owing to differences in intercalation of the DNA-binding dye ethidium bromide and resulting change in buoyant density, covalently closed circular plasmid DNA (cccDNA) is separated from the nicked, relaxed open and linear forms. The DNA species band at different positions within an ultracentrifuge tube and can be isolated through recovered fractions. Although this technique is still held as the golden shrine of DNA purifications due to the purity of recovered DNA, it relies on a 24 to 48 hour centrifugation step to separate the different DNA forms. More recently, this slow and tedious technique has given way to the more popular rapid mini-isolation methods which can be performed in less than an hour (2, 3). Today, the most widely used plasmid mini-prep purification techniques include quick spin filtration combined with binding to ion exchange resins or silica particles.

The simplest method for isolation of plasmid DNA employs the disruption of bacterial cells with phenol/chloroform. Simultaneous extraction of the crude bacterial lysate with organic substances followed by a quick centrifugation step allows the separation of denatured proteins, cell membrane components, and chromosomal DNA, while the plasmid DNA remains in the aqueous phase. Many single-tube procedures have been devised (4-9), some of which have been shown to produce DNA of sequencing quality (10-14). Unfortunately, this method requires the use of caustic chemicals, prohibiting its routine use without protective clothing, gloves, eyewear, and a fume hood.

Another common method involves mixing of bacterial suspensions with a solution of lysozyme and the non-ionic surfactant Triton® X-100. Upon boiling, the lysed bacteria spill forth cellular material which forms an insoluble clot of denatured components that can be removed by centrifugation. The pellet of debris is either removed with a toothpick, or the supernatant containing the plasmid DNA is transferred to a clean tube for further manipulations (15-19).

By far, the most commonly used technique is the alkaline lysis method. Cells are disrupted with the anionic detergent sodium dodecyl sulfate (SDS) in the presence of the alkali sodium hydroxide, causing the denaturation and precipitation of bacterial proteins and other cellular debris (20, 21). Irreversible denaturation of linear chromosomal DNA occurs under alkaline conditions. Renaturation of plasmid cccDNA by neutralization with acid occurs more efficiently than renaturation of linear chromosomal DNA due to the close association of the supercoiled strands, allowing the plasmids to remain soluble in the supernatant. The lysate may then be cleared of debris by centrifugation. Many different mini-prep protocols have been developed, all of which are modifications of this method (22-26). While alkaline lysis has been extensively used for Gram-negative bacteria, purification of plasmid DNA from Gram-positive species, including *Bacillus subtilis* and *Staphylococcus aureus*, has also been done (27).

Some techniques are based on the selective precipitation or salting out of bacterial proteins, chromosomal DNA, and plasmids from cell lysates (28), while others utilize detergents which allow proteins and larger polysaccharides to remain in solution during the precipitation of lower molecular weight nucleic acids. For example, cationic alkyltrimethylammonium halides such as cetyltrimethylammonium bromide (CTAB), cetyltrimethylammonium chloride (CTAC), and cetylpyridinium chloride (CPC) precipitate nucleic acids and acid polysaccharides under low salt conditions. The DNA is pelleted by centrifugation and washed with a high salt buffer (29-31).

Purification of Plasmid DNA

For subcloning purposes, separations may be done by electrophoresis through an agarose gel and the DNA band of interest, either the entire cloning vector DNA or a restriction fragment, can be extracted directly from a gel slice. Several clever techniques for the rapid recovery and purification of DNA from agarose gels have recently been developed. However, this method can sometimes be time consuming and the DNA may not be sufficiently pure for subsequent enzyme reactions - some enzymes are sensitive to small contaminating impurities which may co-elute with the DNA. For example, the use of ß-agarase for the enzymatic destruction of agarose polymers and the subsequent precipitation of nucleic acids has resulted in the inability to amplify extracted DNA by the polymerase chain reaction (PCR; 32).

Another molecular size fractionation technique is that of gel filtration through a column of Sephacryl S-1000, a cross-linked co-polymer of allyl dextran and N, N'-methylenebis(acrylamide), or Ultrogel A2. This alternative molecular sieve method uses flow through separation based on size exclusion to resolve cccDNA from the other forms (33-35).

Ion exchange chromatography resins such as RPC-5 or RPC-5 ANALOG, commonly known as the nucleic acid chromatography system (NACS^{TM}) sold by Life Technologies, Inc., Gaithersburg, MD, USA, may be used for analytical and preparative purification of nucleic acids. Due to the attraction of negatively charged phosphate groups to cationic quaternary amines under low salt conditions, nucleic acids bind to the resin particles, usually composed of polychlorotrifluoroethylene coated with alkyltrimethylammonium chloride. The DNA is eluted with increasing salt by charge competition. Many of the commercially available kits used for mini-prep plasmid purifications contain comparable resins (36, 37).

Hydroxyapatite (HAP) chromatography has also been used extensively for the purification of nucleic acids. Interactions of calcium ions on the surface of HAP (calcium phosphate hydroxide) with high molecular weight nucleic acids trap the DNA onto the matrix while the lower molecular weight extrachromosomal cccDNA flows through (38).

The use of diatomaceous earth, the fossilized cellular remains of unicellular algae (diatoms), and silica particles from other sources has gained great attention recently. In the presence of a chaotropic buffer composed of highly positively charged molecules, usually guanidine (aminomethanamidine), DNA molecules have very strong affinity for siliceous materials. Fine particles of diatomaceous earth and other amorphous forms of silicon dioxide possess many fine pores and, having a high surface-to-volume ratio, are therefore extremely absorbent. Although the exact nature of the electrostatic interaction between the DNA molecules and the silicon dioxide is not known, the association is strong enough to hold small fragments of DNA during vigorous flow washing (39-43). Smaller DNA molecules bond more tightly and are more difficult to elute. Also, larger molecules may bind to more than one particle and be sheared during manipulations of the matrix with bound DNA. Due to this trade-off of properties, it is recommended that the optimal size of DNA to be purified using this method is between approximately 100 base pairs and 6 kilobase-pairs of length, although double-stranded DNA as large as 48 kb has been recovered from clinical specimens with 50% lower efficiency (40).

Many resins currently available are microporous substances treated with silicifying agents. In addition, paramagnetic particles have been coated with a silicide, making them amenable to magnetic separation techniques after DNA is bound. Others have used crystalline silicon dioxide fragments from other sources such as crushed flint glass, ground glass filters, and silica particles (44-46).

Future Trends in DNA Purification

The use of high-end equipment such as a capillary electrophoresis system or HPLC for the purification of cccDNA is gaining popularity. Due to the low quantity of cccDNA recovered, and the expense and time required for setting up and maintaining a system in a low-funded lab, however, these methods are not widely used in molecular biology laboratories. This has probably prohibited the exploration of many potentially useful techniques. Simple procedures requiring less expensive materials allow the average lab worker to complete many plasmid isolations in a single day. The future therefore looks quite bright for the further development of faster methods and new binding resins to be used in mini-prep procedures.

The trend of DNA purification techniques toward commercial kit reagents does have its consequences. Many companies now sell excellent quality controlled binding resins and the materials provided are becoming more reliable and quality controlled. Unfortunately, problems may be encountered if the proprietary binding matrix is changed or removed from the market. For example, many scientists were faced with changing their experimental conditions when the undisclosed binding matrix in a widely used mini-prep kit was recently discontinued (47, 48). In the future, companies providing these materials may consider revealing more details concerning the manufacturing of their products, or they may suffer setbacks within the competitive biotechnology market place.

Below is a mini-prep procedure I have used for routine plasmid isolations.

Protocols

Mini-prep Method

1. Grow overnight cultures of *E. coli* cells containing the recombinant plasmids to be screened in 3-5 ml of selective broth.

2. Pellet 2-4 ml of the cells in a 2.0 ml eppendorf tube. This can be done by spinning 2.0 ml of culture twice for 1 minute in the eppendorf centrifuge or pelleting the entire culture in a clinical centrifuge for 5 minutes.

3. Resuspend the bacterial cell pellet in 100 μl of 25 mM Tris, 10 mM EDTA, 100 μg/ml RNaseA, pH 8.0

4. Using a micropipette, squirt in 200 μl of 0.15 M NaOH, 1.0% SDS pre-warmed to 65 °C on top of the cells without mixing.

5. Add 150 μl of 3 M sodium acetate pH 4.8 to each tube and mix by flicking the tube approximately ten times with your fingernail.

6. Spin for 5-10 minutes in the eppendorf centrifuge at 4 °C.

7. Mix in 1.0 ml of binding solution containing the binding matrix, composed of either diatomaceous earth purchased from Sigma Chemical Co. or Aldrich Chemical Co., St. Louis, MO, USA, or other silica particle resin such as GLASSMILK® from BIO101, La Jolla, CA, USA. While avoiding the white precipitate on the bottom, immediately transfer the solution to a pre-made tip column placed within a vacuum manifold apparatus (see sections below for details).

8. Apply the vacuum to concentrate the binding matrix onto the filter within the column.

9. Wash the trapped matrix twice by rinsing with 1.0 ml of 10 mM Tris; 100 mM NaCl; 2.5 mM EDTA; 55% v/v Ethanol; pH 7.5 with the vacuum applied.

10. Allow the packed matrix to completely dry by applying the vacuum for a further 5 to 10 minutes.

11. Remove each tip and slice off the end just below the filter with a razor blade, leaving no space between the filter and the end. Place the top portion of this, which contains the trapped matrix, inside a 1.6 ml eppendorf centrifuge tube. It is important that a 1.6 ml conical centrifuge tube be used at this point since the inserted cutoff tip will leave a 20 µl volume of dead space as a collection reservoir in the bottom of the conical tube. These may be purchased from various supply companies.

12. Elute the DNA bound to the resin by pipetting 30 µl of sterile distilled water on top of the matrix.

13. The DNA solution is collected by centrifugation for 2 minutes in the eppendorf centrifuge at top speed at room temperature. Remove the plastic tip from the tube with forceps. The liquid still adhering to the tip can be released by smartly tapping the cutoff tip onto the side of the tube.

Preparation of the Binding Matrix and Binding Solution

Method 1 - Use of Diatomaceous Earth as Binding Matrix

The binding matrix is prepared by mixing 3.0 grams of diatomaceous earth with 30 ml of sterile deionized water to give a solution of 100 mg/ml. This is allowed to settle for two to three hours, and the upper liquid phase discarded. The settled particulate matter is resuspended in 6.0 ml of sterile deionized water to give a final stock solution of approximately 150 mg/ml w/v (density of 1.130 to 1.150 g per ml). To prepare the binding matrix, 2.0 ml of this stock solution is added to 50.0 ml of chaotropic binding solution composed of either 7 M guanidine HCl in 50 mM Tris; 20 mM EDTA; pH 7.0, or 6 M guanidine thiocyanate in 50 mM Tris; 20 mM EDTA; pH 7.0.

An alternative method is to make the chaotropic binding solution separately from the binding matrix and add 30 µl of binding matrix stock to the mixed lysate. In this case, the matrix may be prepared by washing a pellet from 1.0 ml of the diatomaceous earth stock solution (150 mg/ml) several times with TE buffer (25 mM Tris; 10 mM EDTA; pH 8.0), before resuspending it in TE for a final concentration of approximately 10 mg/ml, or purchased as Prep-A-Gene® binding matrix from Bio-Rad Laboratories, Hercules, CA, USA.

Method 2 - Use of Ground Glass Silica Particles as Binding Matrix

Silica powder GLASSMILK® may be obtained from BIO101, La Jolla, CA, USA, Celite® from Sigma Chemical Co. or Aldrich Chemical Co., St. Louis, MO, USA, or 325 mesh powdered flint glass fines from Cutter Ceramics, Beltsville, MD, USA. The binding matrix is prepared as for the diatomaceous earth by allowing a suspension to settle for approximately 2 to 3 hours. If impure forms of glass are to be used, the settled particles should be mixed with an equal volume of concentrated nitric acid and boiled in a fume hood. After cooling, the centrifuged pellet of glass powder fines is washed extensively with TE buffer and resuspended to give a final concentration of 10 mg/ml. If silica particles are used, the binding solution should be sodium iodide (908 g/l) saturated with sodium sulfite (15 g/l) added as an antioxidant in 20 mM Tris pH 7.5, or 6 M sodium perchlorate, 50 mM Tris, 10 mM EDTA pH 8.0 (49).

Use of Filter Tips

Although many companies sell filter units that can be used for this mini-prep method, I have found that filtered micropipette tips work very well. To prepare the filter column, pipette 100-200 µl of SigmaCote® silanizing agent through the filter of an aerosol resistant tip (ART®) a few times and allow it to air dry. Push the filter insert down snugly within the tip by using the small end of another pipette tip. To construct the column, fit the syringe barrel of a Becton-Dickinson 1/2 cc U-100 Insulin syringe less the needle, or a 1.0 ml eppendorf pipette tip or similar into the wide end of the silanized filter tip and place this on a vacuum manifold.

It may be important which type of ART® tips can be used since these are typically used to eliminate contaminants for PCR and some, being composed of various hygroscopic materials, are designed to trap moisture. I have routinely used 30-µl Integrity® tips purchased from Matrix Technologies Corporation, Lowell, MA, USA, which contain a 4 mm wide hydrophobic filter fitted directly in the middle of the tip. The maximum capacity of the ART® filter tip was found to be about 30 to 40 µl of concentrated diatomaceous earth resin matrix, or 10-20 µl of silica particle binding matrix. Any more slowed down the vacuum washing steps considerably and caused blockage of the column.

Construction and Use of the Vacuum Manifold

Each column is placed into a PigLet™ universal vacuum manifold purchased from Molecular Bio-Products, San Diego, CA, USA. Use of a commercial manifold is not necessary, however, and an alternative way is to make a homemade system constructed from a discarded 200 µl pipette tip box. To make the manifold, bore a hole within one side of the box just large enough for a nipple fitting. The entire box can be wrapped in parafilm to seal off any cracks or unused holes and a vacuum line hose attached via the nipple. I recommend that a collection trap such as a side arm flask be placed in between the manifold and the vacuum source. The tip columns are then inserted by piercing them through the parafilm covering the holes in the vacuum box system.

Purification of DNA Without a Vacuum Manifold

It is possible to do this procedure without the use of a vacuum manifold or filter devices. Instead, the mixture of cleared lysate, silica particles, and binding solution is transferred to a clean eppendorf tube and centrifuged for 10-15 seconds at top speed. The pellet is then resuspended in 1.0 ml of washing buffer by pipetting and centrifuged again. After two washes, the pellet is dried under vacuum for 5 minutes and the DNA eluted by resuspending it in sterile distilled water. The mixture is again centrifuged and the aqueous DNA solution is pipetted off without disturbing the silica particles. If this method is used, it must be kept in mind that the yield will be lower for larger DNA molecules due to shearing forces.

Scaling-up the Protocol

This method can be scaled up for the recovery and purification of larger amounts of plasmid DNA. Larger syringe barrels may be prepared by plugging the hole with a small amount of siliconized polyallomer aquarium fluff (50). To create a paper filter which is used as the support trap for the binding resin, strips of Whatman® 3MM filter paper can be soaked in TE buffer and mixed by shaking into a slurry, or macerated within a Waring blendor (51). A small portion of the paper slurry is spooned into each syringe such that the passage of liquid is not blocked, but that the binding matrix is trapped on top of the newly formed paper filter during the washing steps. Some experimentation may be needed to determine the correct amount of slurry for each type of syringe and binding material. After the drying step, the DNA can be recovered by adding the appropriate amount of sterile distilled water and centrifuging the syringe within a clinical centrifuge tube fitted into a standard swinging bucket rotor to collect the eluate. At this point, the purified plasmid DNA may be concentrated through precipitation by making the DNA solution 2.5 M ammonium acetate (ie. adding $1/4^{th}$ volume of 10 M stock solution) and 2.5 volumes of absolute ethanol at room temperature. The DNA is collected by centrifuging at 16,000 x g for 15 minutes and the pellet washed several times with 70% ethanol to remove any remaining salts (52).

Purification of Plasmids from Nonviable Glycerol Stocks or Bacterial Colonies

There has been some interest in the recovery of plasmid DNA from nonviable bacteria. Transforming competent *E. coli* cells with the DNA may allow the recovery of a lost recombinant clone in times of desperation (53). Using the method outlined here with a few modifications, I have successfully purified plasmid DNA from glycerol stocks containing nonviable bacteria and was able to recover several seemingly lost clones. The bacteria are pelleted and washed with 1.0 ml of 1 M NaCl and resuspended in TE buffer prior to the lysis step. After addition of the lysis solution the suspensions appear to be more completely lysed. The reason for this is unclear; however, it has been noted that bacterial colonies stored on agar plates at 4 °C for some time produce a substantial amount of exopolysaccharides which may interfere with lysis. The protective coating may be reduced or removed, or the wash may cause the cells to become more sensitive to osmotic shock (54). In any case, this wash step has been demonstrated to be an effective means of lysing bacteria which are normally recalcitrant to this method (55, 56). The entire DNA sample recovered should be added to highly competent cells and the standard procedure for transformation previously described should be followed (57).

Discussion

Using the method outlined in the protocols section, approximately 1-5 µg of purified plasmid DNA in 10-15 µl of solution has been routinely recovered from 12 to 36 mini-preps within 30 minutes, excluding the time needed for bacterial growth. Yield will depend upon the particular plasmid and host strain used. Restriction digestions and agarose gel electrophoresis showed the DNA recovered to be as clean as that obtained by other mini-prep procedures using binding resins and the DNA should be of adequate quality for sequence analysis by the dideoxy termination method. Some modifications to previous methods have been made to produce higher quality DNA. These include (a) the lowering of NaOH concentration from 0.2 to 0.15 M which may reduce damage incurred onto the DNA during exposure to alkaline conditions, and (b) the use of a hot lysis solution allowing for complete lysis upon contact. This allows the immediate neutralization of the alkali and a significant reduction of the time DNA is exposed to denaturing conditions. Other hot alkaline methods require a longer incubation time of 30 minutes (58).

Each of the methods for isolation of plasmid DNA from bacteria reviewed here has its advantages and drawbacks. The method described within this chapter avoids some of the other problems found with DNA purified by other single-tube isolation procedures which utilize organic extractions with phenol and chloroform to denature proteins. While the added expense involved with transferring solutions between tubes is avoided by single-tube mini-prep protocols, the handling of caustic, hazardous, and toxic materials such as phenol and chloroform is required.

On the other hand, there have been reports that omitting the organic extraction step can yield lesser quality DNA. It has been suggested that degradation of recovered DNA is most likely due to endonucleases that remain active during the purification steps. It has been recommended that *endA*$^-$ strains of *E. coli* be used for maintaining the plasmid constructs (59). Since the purification protocol described in this chapter optionally includes guanidine thiocyanate, a nuclease inhibitor, this extra precaution may not be necessary.

Trademarks: ART® (aerosol resistant tips) is a registered trademark of Continental Laboratory Products. Celite® is a registered trademark of Manville Service Corporation. GLASSMILK® is a registered trademark of BIO101, Inc. Integrity® is a registered trademark of Matrix Technologies Corporation. NACS™ (nucleic acid chromatography system) is a trademark of Life Technologies. PigLet™ is a trademark of Molecular Bio-Products, Inc. Prep-A-Gene® is a registered trademark of Bio-Rad Laboratories. SigmaCote® is a registered trademark of Sigma Chemical Co. Triton® is a registered trademark of Union Carbide Chemicals and Plastics Company, Inc. Whatman® is a registered trademark of Whatman Paper Ltd.

Acknowledgements

I would like to thank the following people for sharing their experiences, unpublished results, and posted protocols through the BIOSCI supported newsgroup bionet.molbio.methds-reagnts:
C.T.J. Chan [cchan@uk.ac.crc]
Ravi R. Iyer [rriyer@unix1.circ.gwu.edu]
Bill Melchior [wmelchior@ntbtox.nctr.fda.gov]
Chia Jin Ngee [mcblab47@leonis.nus.sg]
Doug Rhoads [drhoads@mercury.uark.edu]
Bruce Roe [broe@aardvark.ucs.uoknor.edu]
Claudia A. Sutton [cas9@cornell.edu]

References

1. Clewell, D. B. and Helinski, D. R. 1969. Supercoiled circular DNA-protein complex in *Escherichia coli:* purification and induced conversion to an open circular DNA form. Proc. Natl. Acad. Sci. USA. 62: 1159-1166.
2. Griffith, O. M. 1988. Large-scale isolation of plasmid DNA using high speed centrifugation methods. BioTechniques. 6: 725-727.
3. Stemmer, W. P. C. 1991. A 20-minute ethidium bromide/high-salt extraction protocol for plasmid DNA. BioTechniques. 10: 726.
4. Alter, D. C. and Subramanian, K. N. 1989. A one step, quick step, mini prep. BioTechniques. 7: 456-458.
5. Serghini, M. A., Ritzenthaler, C. and Pinck, L. 1989. A rapid and efficient 'miniprep' for isolation of plasmid DNA. Nucl. Acids Res. 17: 3604.
6. He, M., Wilde, A. and Kaderbhai, M. A. 1990. A simple single-step procedure for small-scale preparation of *Escherichia coli* plasmids. Nucl. Acids Res. 18: 1660.
7. Chowdhury, K. 1991. One step 'miniprep' method for the isolation of plasmid DNA. Nucl. Acids Res. 19: 2792.
8. Akada, R. 1994. Quick-check method to test the size of *Escherichia coli* plasmids. BioTechniques. 17: 58.
9 Kurien, B. T. and Scofield, R. H. 1994. A short, small-scale preparation of plasmid DNA from *E. coli* using organic solvents. B.R.L. Focus. 16: 113.
10. Johnson, K. R. 1990. A small-scale plasmid preparation yielding DNA suitable for double-stranded sequencing and *in vitro* transcription. Anal. Biochem. 190: 170-174.
11. Kibenge, F. S. B., Dybing, J. K. and McKenna, P. 1991. Rapid procedure for large-scale isolation of plasmid DNA. BioTechniques. 11: 65-67.
12. Goode, B. L. and Feinstein, S. C. 1992. "Speedprep" purification of template for double-stranded DNA sequencing. BioTechniques. 12: 374-375.
13. Yie, Y., Wei, Z. and Tien, P. 1993. A simplified and reliable protocol for plasmid DNA sequencing: fast miniprep and denaturation. Nucl. Acids Res. 21: 361.
14. Tarczynski, M. C., Meyer, W. J., Min, J. J., Wood, K. A. and Hellwig, R. J. 1994. Two-minute miniprep method for plasmid DNA isolation. BioTechniques. 16: 514-519.

15. Holmes, D. S. and Quigley, M. 1981. A rapid boiling method for the preparation of bacterial plasmids. Anal. Biochem. 114: 193-197.
16. Wang, L.-M., Weber, D. K., Johnson, T. and Sakaguchi, A. Y. 1988. Supercoil sequencing using unpurified templates produced by rapid boiling. BioTechniques. 6: 839-843.
17. Liszewski, M. K., Kumar, V. and Atkinson, J. P. 1989. "Midi-prep" isolation of plasmid DNA in less than two hours for sequencing, subcloning and hybridizations. BioTechniques. 7: 1079-1081.
18. Berghammer, H. and Auer, B. 1993. "Easy-Preps": fast and easy plasmid minipreparation for analysis of recombinant clones in *E. coli*. BioTechniques. 14: 524-528.
19. Hultner, M. L. and Cleaver, J. E. 1994. A bacterial plasmid DNA miniprep using microwave lysis. BioTechniques. 16: 990-994.
20. Birnboim, H. C. and Doly, J. 1979. A rapid alkaline extraction procedure for screening recombinant plasmid DNA. Nucl. Acids Res. 7: 1513-1523.
21. Birnboim, H. C. 1983. A rapid alkaline extraction method for the isolation of plasmid DNA. Meth. Enzym. 100: 243-255.
22. Morelle, G. 1989. A plasmid extraction procedure on a miniprep scale. B.R.L. Focus 11: 7-8.
23. Zhou, C., Yang, Y. and Jong, A. Y. 1990. Mini-prep in ten minutes. BioTechniques. 8: 172-173.
24. Jones, D. S. C. and Schofield, J. P. 1990. A rapid method for isolating high quality plasmid DNA suitable for DNA sequencing. Nucl. Acids Res. 18: 7463-7464.
25. Xiang, C., Wang, H., Shiel, P., Berger, P. and Guerra, D. J. 1994. A modified alkaline lysis miniprep protocol using a single microcentrifuge tube. BioTechniques. 17: 30-32.
26. Wang, L.-F., Voysey, R. and Yu, M. 1994. Simplified large-scale alkaline lysis preparation of plasmid DNA with minimal use of phenol. BioTechniques. 17: 26-28.
27. Voskuil, M. I. and Chambliss, G. H. 1993. Rapid isolation and sequencing of purified plasmid DNA from *Bacillus subtilis*. Appl. Environ. Microbiol. 59: 1138-1142.
28. Laitinen, J., Samarut, J. and Hölttä, E. 1994. A nontoxic and versatile protein salting-out method for isolation of DNA. BioTechniques. 17: 316-322.
29. Del Sal, G., Manfioletti, G. and Schneider, C. 1989. The CTAB-DNA precipitation method: a common mini-scale preparation of template DNA from phagemids, phages or plasmids suitable for sequencing. BioTechniques. 7: 514-520.
30. Ishaq, M., Wolf, B. and Ritter, C. 1990. Large-scale isolation of plasmid DNA using cetyltrimethylammonium bromide. BioTechniques. 9: 19-24.
31. Chang, P. K. and Natori, M. 1994. 'One for all scales' methods for plasmid, phagemid and bacteriophage DNAs preparation. J. Biotechnol. 36: 247-251.
32. Hengen, P. N. 1994. Methods and reagents - recovering DNA from agarose gels. Trends in Biochem. Sci. 19: 388-389.
33. Micard, D., Sobrier, M. L., Couderc, J. L. and Dastugue, B. 1985. Purification of RNA-free plasmid DNA using alkaline extraction followed by Ultrogel A2 column chromatography. Anal. Biochem. 148: 121-126.
34. Gómez-Márquez, J., Freire, M. and Segade, F. 1987. A simple procedure for large-scale purification of plasmid DNA. Gene 54: 255-259.
35. Raymond, G. J., Bryant III, P. K., Nelson, A. and Johnson, J. D. 1988. Large-scale isolation of covalently closed circular DNA using gel filtration chromatography. Anal. Biochem. 173: 125-133.

36. Best, A. N., Allison, D. P. and Novelli, G. D. 1981. Purification of supercoiled DNA of plasmid Col E1 by RPC-5 chromatography. Anal. Biochem. 114: 235-243.
37. Thompson, J. A., Blakesley, R. W., Doran, K., Hough, C. J. and Wells, R. D. 1983. Purification of nucleic acids by RPC-5 ANALOG chromatography: peristaltic and gravity-flow applications. Meth. Enzym. 100: 368-399.
38. Shoyab, M. and Sen, A. 1979. The isolation of extrachromosomal DNA by hydroxyapatite chromatography. Meth. Enzym. 68: 199-206.
39. Hall, R. H. 1967. Partition chromatography of nucleic acid components (isolation of the minor nucleosides). Meth. Enzym. 12A: 305-315.
40. Boom, R., Sol, C. J. A., Salimans, M. M. M., Jansen, C. L., Wertheim-van Dillen, P. M. E. and van der Noordaa, J. 1990. Rapid and simple method for purification of nucleic acids. J. Clin. Microbiol. 28: 495-503.
41. Carter, M. J. and Milton, I. D. 1993. An inexpensive and simple method for DNA purifications on silica particles. Nucleic Acids Res. 21: 1044.
42. Willis, E. H., Mardis, E. R., Jones, W. L. and Little, M. C. 1990. Prep-A-Gene^{TM}: A superior matrix for the purification of DNA and DNA fragments. BioTechniques. 9: 92-99.
43. Pan, H.-Q., Wang, Y.-P., Chissoe, S.L., Bodenteich, A., Wang, Z., Iyer, K., Clifton, S.W., Crabtree, J.S. and Roe, B.A. 1994. The complete nucleotide sequences of the SacBII Kan domain of the P1 pAD10-SacBII cloning vector and three cosmid cloning vectors: pTCF, svPHEP, and LAWRIST16. Genet. Anal. Tech. Appl. 11: 181-186.
44. Vogelstein, B. and Gillespie, D. 1979. Preparative and analytical purification of DNA from agarose. Proc. Natl. Acad. Sci. USA. 76: 615-619.
45. Marko, M. A., Chipperfield, R. and Birnboim, H. C. 1982. A procedure for the large-scale isolation of highly purified plasmid DNA using alkaline extraction and binding to glass powder. Anal. Biochem. 121: 382-387.
46. Sparks, R. B. and Elder, J. H. 1983. A simple and rapid procedure for the purification of plasmid DNA using reverse-phase C18 silica beads. Anal. Biochem. 135: 345-348.
47. Hengen, P. N. 1994. Methods and reagents - kit wars. Trends in Biochem. Sci. 19: 46-47.
48. Hengen, P. N. 1994. Methods and reagents - on the magic of mini-preps.... Trends in Biochem. Sci. 19: 182-183.
49. Hengen, P. N. 1994. Frequently Asked Question (FAQ) list for bionet.molbio.methds-reagnts version number 01.13.10.1994 available via anonymous FTP from ftp.ncifcrf.gov as file pub/methods/FAQlist or upon request by e-mail to pnh@ncifcrf.gov
50. Levine, R. A. 1994. Aquarium filter floss: an alternative to silanized glass wool as a porous support matrix. BioTechniques. 17: 67.
51. Chuang, S.-E. and Blattner, F. R. 1994. Ultrafast DNA recovery from agarose by centrifugation through a paper slurry. BioTechniques. 17: 634-636.
52. Crouse, J. and Amorese, D. 1987. Ethanol precipitation: ammonium acetate as an alternative to sodium acetate. B.R.L. Focus. 9(2): 3-5.
53. Shuldiner, A. R. and Tanner, K. 1992. Recovery of plasmid DNA from nonviable bacterial colonies and cultures. BioTechniques. 12: 66.
54. Schwinghamer, E. A. 1980. A method for improved lysis of some Gram-negative bacteria. FEMS Microbiol. Lett. 7: 157-162.

55. Dénarié, J., Boistard, P., Casse-Delbart, F., Atherly, A. G., Berry, J. O. and Russell, P. 1981. Indigenous plasmids of *Rhizobium*. In: International Review of Cytology, Supplement 13. Biology of the *Rhizobiaceae*. K. L. Giles and A. G. Atherly, eds. Academic Press, New York. p. 225-246.
56. Kuykendall, L. D. and Hengen, P. N. 1988. Microbial genetics of legume root nodulation and nitrogen fixation. In: Biological Nitrogen Fixation - Recent Developments. N. S. Subba Rao, ed. Oxford and IBH Publishing Co., New Delhi, India. p. 71-111.
57. Hengen, P. N. and Iyer, V. N. 1992. DNA cassettes containing the origin of transfer (*ori*T) of two broad-host-range transfer systems. BioTechniques. 13: 57-62.
58. Musich, P. R. and Chu, W. 1993. A hot alkaline plasmid DNA miniprep method for automated DNA sequencing protocols. BioTechniques. 14: 958-960.
59. Taylor, R. G., Walker, D. C. and McInnes, R. R. 1993. *E. coli* host strains significantly affect the quality of small scale plasmid DNA preparations used for sequencing. Nucl. Acids Res. 21: 1677-1678.

From: *Molecular Biology: Current Innovations and Future Trends.*
ISBN 1-898486-01-8 ©1995 Horizon Scientific Press, Wymondham, U.K.

4

GEL ELECTROPHORESIS OF DNA AND PROTEINS: RECENT ADVANCES IN THEORY AND PRACTICAL APPLICATIONS

Branko Kozulić

Abstract

In order to explain the mechanism of electrophoretic migration of macromolecules through gels, it is necessary to define the spaces occupied by the migrating molecules and to describe how the molecules pass through these spaces. Some models of gel electrophoresis consider that the spaces (pores) exist regardless of the presence of the migrating macromolecules, whereas others propose that the migrating molecules influence or create these spaces. The gel polymers are treated as fixed obstacles or displaceable, elastic chains. The present models of gel electrophoresis are compared and their predictions correlated to the published experimental data. Recent progress in the practical applications of gel electrophoresis is briefly reviewed, with the emphasis on emerging techniques and on the useful, but less frequently employed procedures.

Recent Advances in the Theory of Gel Electrophoresis

Gel electrophoresis is currently the most widely used method for the analysis of biological macromolecules, including proteins and nucleic acids. In spite of the considerable scientific work published on the theory and applications of gel electrophoresis, the field remains a subject of intense investigation. Several reviews have been published recently (1-5). One of the goals of this article is to critically survey novel theoretical advances in gel electrophoresis.

The first integral theory of gel electrophoresis is known as the extended Ogston model (6), which is based on the analysis of the spaces available to a spherical object placed in a suspension of straight, long fibers (7). According to this model, mobility is a function of the fraction of gel pores available to the migrating molecule, which depends on the probability of no contact with the gel fibers. The probability is related to the size of the molecule, the thickness of gel polymers and their concentration (reviewed in 8). The model was established using mobility data for native proteins and it is consistent with many experimental results (6, 8, 9). In particular, the slopes of the straight lines obtained in the plots of gel concentration versus logarithm of mobility (Ferguson plots) were dependent on protein size. The model assumes that the migrating molecules have a globular shape, which applies to most native proteins. However, since protein-SDS complexes are elongated (10) and nucleic acid fragments are thread-like molecules, the extended Ogston model could be applied to such macromolecules only by assuming that elongated molecules were spheres defined by their radius of gyration (6).

The finding by Southern (11) that the mobilities of DNA fragments are proportional to the reciprocal of their sizes, led to the development of the reptation (snake-like movement) model of DNA gel electrophoresis (12-15). The reptation model considers that the DNA migration path is constricted by gel fibers so that a DNA molecule migrates essentially through holes which are connected to form a "tube". The model is based on de Gennes' theory of motion of a polymeric molecule in the presence of fixed obstacles (16). The mobilities of large DNA molecules and of smaller DNA molecules at high electric field strengths, will approach a constant value. At that point, the resolution of DNA molecules of different sizes will be lost. This was experimentally observed for long DNA molecules in agarose gels (11, 17).

Both the extended Ogston and reptation models of gel electrophoresis consider that the migrating molecule moves through a continuous series of gel pores, which may have a particular size distribution (see figure 17 in reference 8). A DNA molecule should reptate only when the gel pore radius is smaller than the radius of gyration of the migrating molecule. In several recent reports the reptation model was therefore proposed to apply only to long DNA molecules and the extended Ogston model to apply only to proteins and short DNA molecules (2, 18, 19).

The extended Ogston model predicts well the electrophoretic behaviour of proteins run in cross-linked polyacrylamide gels of different concentrations, with only minor exceptions (8). The reptation model is able to explain the behaviour of large DNA molecules in free solution (20, 21).

Schwartz and Cantor, based on predictions of the reptation model, introduced pulsed field gel electrophoresis (22), which has revolutionized the separation of long DNA molecules (see Chapter 5). As the experimental data from pulsed field experiments accumulated, the initial reptation model underwent refinements (23-32), which were greatly augmented by direct observations of individual DNA molecules in unidirectional and pulsed electric fields (33-37). The behaviour of large DNA molecules during gel electrophoresis is more complex than initially thought, resembling more the movement of a caterpillar than a snake (33-40).

To interpret experimental results using the extended Ogston or the reptation models, it is necessary to know the pore sizes of a gel. The values for the pore sizes of the commonly used polyacrylamide and agarose gels are still a matter of dispute, however, and are dependent on the analytical method used (41). Even with a single method, polyacrylamide gel electrophoresis, pore sizes estimated with DNA were one order of magnitude higher than those estimated with proteins using the extended Ogston model (42, 43). A problem arose when this model was used for interpretation of the results obtained by polyacrylamide and agarose gel electrophoresis of small and medium size DNA molecules. The mobilities could be correlated well to their sizes only after extrapolating the electric field strength to zero (44), which is experimentally a non-real value.

Strong doubts about the validity of the reptation model were raised recently by two of its most prominent authors (45, 46). The major reason was the finding that the scaling power between size and mobility decreased to -3 for long DNA fragments (47). According to the classic reptation model, the scaling power can decrease to -1, but not more (46). Another challenge to the reptation model is the phenomenon of "band inversion", where a large DNA molecule migrates faster than a smaller one in gels run in an unidirectional electric field (48, 49). This contradicts the reptation model, which predicts that mobility decreases inversely with size and approaches a constant value for long DNA molecules (12-15). It also contradicts the extended Ogston model, because the molecule with a larger radius should always migrate more slowly (6). To explain this phenomenon, also called "band scrambling" (50, 51), a "biased reptation model" was proposed (2, 15, 50, 51). According to this model, "band inversion" is related to dynamic, temporary, self-trapping of the migrating molecules in an inverted U-like conformation. A smaller molecule will migrate more slowly if it spends more time in this U-shaped conformation than a larger molecule.

A different kind of problem was encountered during our study of novel matrices for electrophoresis. A branched cross-linker, when mixed with its linear precursor and copolymerized with a monomer, formed a gel that failed to resolve DNA fragments (52). In contrast, resolution of DNA fragments was normal when the gel contained only the starting, linear vinyl-agarose cross-linker (52). This unexpected result, which is inconsistent with either the extended Ogston or reptation models, can be explained by a new model of gel electrophoresis, called the "door-corridor" model (52-55).

According to the door-corridor model, there are no pre-existing pores of finite shape or size in a gel. Migrating macromolecules push the polymers aside, clearing the space they occupy. The macromolecules move in discrete steps and in each step they pass through one gel layer. The essential feature of the door-corridor model is the notion that there are two ways a macromolecule can pass through a gel layer, via a door or via a corridor. Doors are openings formed in the region of a gel layer in which the polymer chains have a high motional freedom. Formation of a door does not affect the polymer chains in other gel layers. Corridors are openings formed in the gel layer where the polymers have a low motional freedom. To form a corridor, the migrating molecule must deform a gel layer until an opening develops at a place where one or more polymer chains end or where the polymer chains are less cross-linked or entangled. The deformation of one gel layer is accompanied by dislocation of some polymer chains in at least one layer above and below. If at the next layer the migrating macromolecule encounters a similar area of low motional freedom, it will again open a corridor. The two corridors may fuse into a single long corridor, spanning several gel layers. To allow for the opening of corridors, the migrating molecule must be able to sufficiently displace the polymers of different gel layers. The branched cross-linker, which linked many polymers from different layers, did not permit a sufficient polymer chain dislocation, resulting in the loss of gel resolving power (52).

The door-corridor model further proposes that the alternative between opening predominantly doors or corridors by a particular migrating macromolecule depends entirely on the balance of two forces. The first force, acting on gel layers, is electrokinetic and it is exercised by all macro-ions moving in the electric field. This force is countered by the resisting force of the polymer chains comprising the gel layer. When the ratio of polymer resistance to electrokinetic force is high, corridors are formed and when this ratio is small, the openings will be predominantly doors. If the two forces are of equal magnitude, then the migrating macromolecule will open doors and corridors in equal ratio.

During gel electrophoresis, the electrokinetic force of the migrating macromolecule is countered by deformable, elastic polymers whose resistance depends on the electrokinetic force itself. To pass through a gel layer, the macromolecule may need to use all of its force, or alternatively, just a fraction of its force may be sufficient to form the openings. The molecules which are sufficiently small to pass between gel polymers are met by little resistance, and they do not need to use their full force to create openings (mostly doors). Large molecules encounter a lot of resistance, but the polymers in a gel layer can resist only up to a certain point. When that point is surpassed, large molecules also do not have to use their entire force. Such molecules create long corridors in the gel layers and this is then accompanied by "band inversion" and loss of gel resolving power (52-55).

According to the door-corridor model, elastic forces govern the migration of macromolecules through gels. Interestingly, friction is not constant per DNA segment, but increases with size. During gel electrophoresis, the migrating molecules are pushed against the gel polymers, which creates a high friction and limits thermally induced diffusion of the migrating molecules. This results in sharp bands even at elevated temperatures (55). The practical importance of the door-corridor model is that it can predict which molecules are optimally resolved in a gel. It also allows a calculation of the DNA net charge per base pair, the electrokinetic force of different DNA molecules, thermal diffusion of the migrating molecules and the resisting forces of the gel polymers (55).

Other models of gel electrophoresis are also known. Here we can only briefly discuss them; a more detailed comparison will be published elsewhere (manuscript in preparation). Bode proposed that finite pores are not necessary and that the mechanism of gel electrophoresis can be explained in terms of polymer viscosity (56-60). Calladine *et al.* (47) suggested that the mechanism of DNA migration in agarose and polyacrylamide gels needs to be explained by two models. The first model was similar to the extended Ogston model and according to the second model, DNA molecules pile against a gel barrier until it breaks or deforms under the resulting force (47).

A model for pulsed field gel electrophoresis of DNA was recently described by Chu (61). A large DNA molecule is pictured as a deformable "bag", whose shape and orientation depends on the strength and direction of the electric field. The model can predict the electrophoretic mobility of megabase DNA fragments, with only minimal assumptions regarding gel properties and DNA conformation (61). Another, similar model has also been described (62).

The related models can be categorized based on several criteria. For example, the extended Ogston and reptation models assume that finite pores exist in electrophoresis gels, so that the migrating molecules follow a pre-existing path. In contrast, finite, pre-existing pores are not necessary according to the model of Bode (56-60), the second model of Calladine *et al.* (47), and the door-corridor model (52-55). According to all models, except the extended Ogston, gel polymers can influence the shape of the migrating molecule. The suggestion that the migrating molecule can create the spaces through which it moves is consistent with the second model of Calladine *et al.* and the door-corridor model.

As we can see, the existing models of gel electrophoresis treat the gel matrix and the migrating molecules in different, often opposing, ways. Yet they all aim at explaining the same mobility data. While each model is good at interpreting a subset of data, some models face major difficulties in explaining all the experimental findings.

"Band inversion" represents the first challenge to the extended Ogston and classic reptation models. The second challenge is the finding that uncross-linked polymer solutions are almost as effective as cross-linked polymerized gels in separating proteins and nucleic acids by capillary electrophoresis (63-67). This is the principal problem for all models relying on migration of macromolecules through gel pores, because in a solution of free uncross-linked polymers finite pores cannot exist (68, 69). The third challenge is the frequent observation that the distance migrated by DNA molecules depends on the amount of DNA in the sample (70-72) and that applying a large sample (overloading) leads to a loss of resolution. A fourth challenge is the important finding that the gel polymers themselves can be directly affected by the electric field (73-76). These findings, as well as other unexpected observations (77-81), can best be explained by the door-corridor model (52-55).

Recent Advances in the Practical Applications of Gel Electrophoresis

Proteins

There are many variations of gel electrophoresis that are used for the analysis of proteins and nucleic acids. Some have reached the status of general acceptance while the others are well established for particular applications.

Separation of proteins on the basis of charge is achieved by isoelectric focusing (IEF), whereas in SDS electrophoresis, proteins are separated according to size. The combination of the two techniques into two-dimensional (2-D) electrophoresis gives a method able to resolve thousands of proteins. All three methods are well standardized and detailed protocols are available (82-86).

Proteins are usually denatured prior to electrophoresis and IEF. While running a gel under denaturing conditions allows a separation either according to size or charge, the information on quaternary structure of oligomeric proteins is lost. It is worth noting that oligomeric proteins can be studied by native gradient gel electrophoresis using steep (3-30 %) gradients, provided that the running time and buffer pH are properly selected (87, 88). This method is particularly suited for studying the assembly of oligomeric proteins (89, 90) and their cross-linking (91, 92). Blue native gel electrophoresis can be also used for the same purpose (93).

Caution is necessary when determining the size or charge of an unknown protein by electrophoresis. Some proteins, in particular glycoproteins and very hydrophobic proteins, show anomalous mobilities, because the first bind less and the latter bind more SDS than the average protein does (84, 94). In IEF, the cyanate present in the urea solution can carbamylate lysine groups in proteins, leading to artifacts (95). Spurious bands may appear due to oxidation of sulfhydryl groups to disulfides by either oxygen from the air or by the oxidizing species left over from gel polymerization (96, 97). The last problem can be easily overcome by alkylation of the -SH groups with, for example, iodoacetamide. An alternative, odorless reagent for reduction of disulfide bridges was proposed recently (98).

The separated proteins are detected with various stains. Most widely used are Coomassie blue dyes and silver (the latter offers about 100-fold higher sensitivity). In our hands the fast, reversible and sensitive zinc staining (99) also works well. Many enzymes can be detected in the gels by activity staining (100). Protein molecules inside a gel can be subjected to various chemical reactions *in situ,* including specific peptide bond cleavage reactions in order to study the homology of the protein species resolved in the gel (90, 101-103).

Blotting of the separated proteins to a membrane is frequently done for specific detection by antibodies or other analyses (104). The proteins can also be blotted during electrophoresis, onto a moving membrane in contact with the gel (105, 106). The identity of a membrane-bound protein can be determined by microsequencing, as reviewed recently (107).

Precast gels are a relatively new development in gel electrophoresis. They are designed to optimize protein resolution and gel reproducibility, while minimizing exposure to toxic gel monomers and significantly reducing the amount of hazardous waste.

DNA

DNA molecules are unique among other macro-ions in that they span a very wide size range, from a few base pairs to millions of base pairs. The contour length of a 6 megabase pair molecule is about 2 mm, which is close to the typical width of a visible band in the agarose gels. The fact that such long molecules can migrate and separate into discrete zones during electrophoresis is amazing. Not surprisingly, the gels must be run under special conditions.

In contrast to proteins, sharp DNA bands can be obtained without the use of a discontinuous buffer system, because the molecules concentrate at the site of their entrance into the gel. Even though double stranded DNA molecules can be analyzed in all formats, submerged (submarine) gel electrophoresis is the format of choice, as it eliminates many problems associated with other modes. With a gel submerged in the running buffer, there is a continuity from one electrode to another, the electric field is steady, leakage cannot occur since the anode and cathode buffer tanks are at the same level, and the application of samples is simple. The classic design of the submerged electrophoresis unit is satisfactory for many uses, and newer apparati offer improvements that increase uniformity of electric field (108, 109), temperature control (108-110) and buffer mixing (108-111).

Over 50% of all DNA samples, including most PCR samples, are analyzed using the submerged gel electrophoresis mode. Gel electrophoresis is preferred because, at a low cost, it gives more information than other methods about the complexity of a DNA sample, and about the size and concentration of the individual components of the sample. The throughput issue, often cited as the limitation of gel electrophoresis, is addressed by the gel modifications which allow simultaneous running of about 100 samples, as described for example in (112).

To estimate the size of a DNA molecule (Sz), one compares its electrophoretic mobility to that of a set of marker DNA fragments of known sizes. A semi-log standard curve is usually constructed, although the reciprocal of mobility plot may be advantageous. Such a plot (Figure 1) shows an excellent linearity. The requirement for knowing the absolute mobilities can be circumvented by taking the migration distances (d) relative to a selected fragment. When that fragment corresponds to the one optimally resolved in the gel, a simple equation of the form $Sz_{unknown} = (d_{opt}/d_{unknown}) \cdot A - B$,

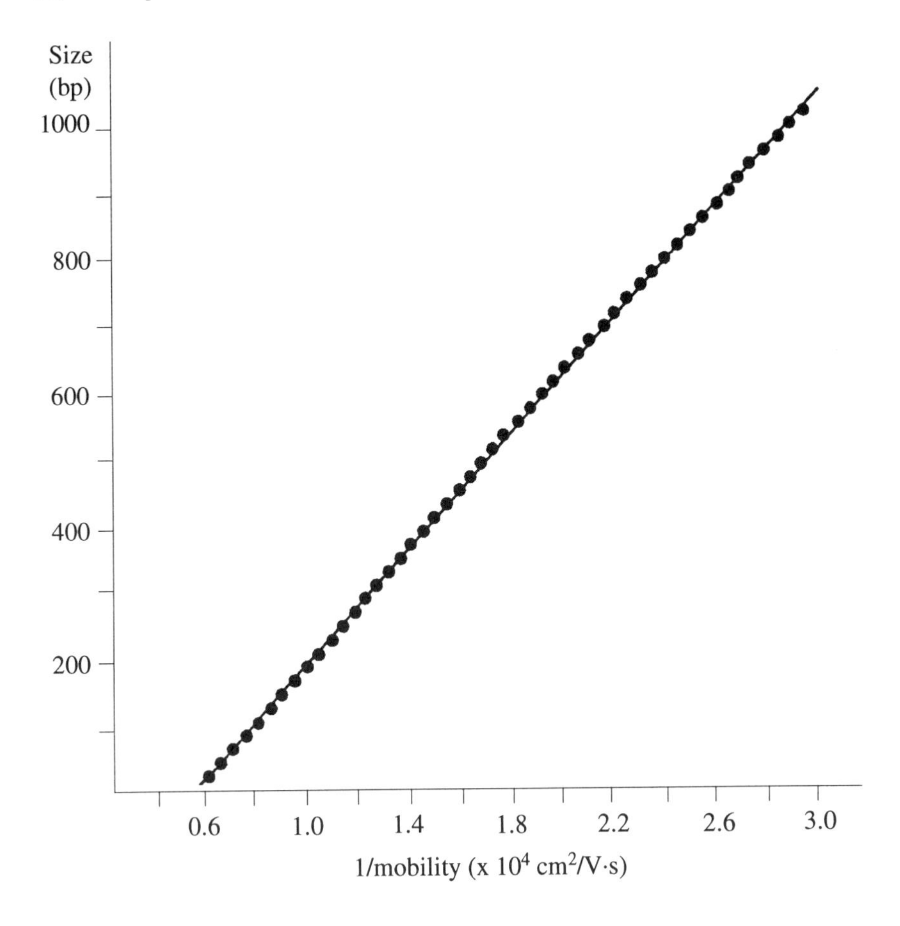

Figure 1. The plot of DNA sizes versus the reciprocal of their mobilities. DNA molecules from the 20 bp ladder (Superladder-low, Gensura) were run simultaneously in three precast 6% poly(NAT) gels at 5 V/cm in the SEA-2000 submerged gel electrophoresis apparatus (Guest Elchrom Scientific). The first gel was run for 2 h 45 min, the second for 4 h 15 min, and the third for 5 h 45 min, in order to allow more precise measurement of migration distances. The mobilities calculated from the three gels differed less than 2%, and the mean values were used for constructing the plot. The gels were equilibrated and run in 80 mM TAE. At lower TAE concentrations, the linearity is lost below about 100 bp (unpublished), and a deviation occurs also above about 2000 bp (53-55).

where A is 2.718/k_{dc} and B is 1/k_{dc}, can be conveniently employed (k_{dc} is gel constant, ref. 52-55). Markers with a large number of closely spaced fragments covering various size ranges (see Figure 2), can replace the need for standard curves for most applications.

What is the resolution limit of double stranded DNA gel electrophoresis? On a mini-gel (about 10 cm long), in the size range from 50 to about 2000 bp, in general two fragments can be resolved if they differ by 2% or more (unpublished observations). Occasionally, smaller differences will be resolved but they are related more to the conformation, i.e. DNA curvature (113), of one of the two fragments than to their size difference. Moreover, blunt-ended fragments may migrate differently from fragments with overhangs (see the 123 bp ladder in Figure 2). With an artificial mix of two restriction digests one may observe an unrealistically high resolution (Figure 2, lane 3). In some cases anomalous migration can be larger than 5% (42, 113, 114).

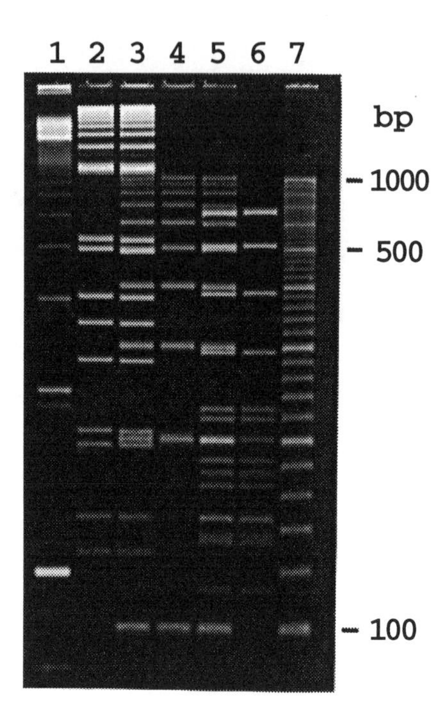

Figure 2. Electrophoresis of dsDNA in a 12% poly(NAT) gel. The gel was run at 12 V/cm for 3 h and 45 min at 20°C. The sample in lane 1 is 123 bp ladder (Life Technologies); 2, 1 kb ladder (Life Technologies); 3, a mixture of the 1 kb and 100 bp ladders; 4, 100 bp ladder (Gensura); 5, a mixture of the 100 bp ladder and pBR322/*Msp* I; 6, pBR322/*Msp* I; and 7, 20 bp ladder. In the 123 bp sample (lane 1), the fragments with overhangs migrate in front of the major bands (less intense bands, best visible is the 246 bp band). In lane 3, the 396 bp band of the 1 kb ladder is resolved from the 400 bp band of the 100 bp ladder. Such a resolution of the 1% difference is exceptional (see text). An anomalous, DNA sequence-dependent migration is evident in lane 3, where the 200 bp band of the 100 bp ladder migrated between the 201 and 220 bp bands of the 1 kb ladder. Some fragments in lane 5 also migrated anomalously.

The excellent resolution of dsDNA is possible in high percentage synthetic gels run in the submerged mode (Figure 2), but only after adjusting the ionic composition inside the gel (115). Otherwise, the separated bands are severely tilted in relation to the migration axis, appearing diffuse when photographed from above (115). The adjustment is necessary also when running a high percentage agarose gel (115, 116).

Several novel techniques, which rely on the resolving power of a gel, have emerged recently. One of them is RAPD (random amplified polymorphic DNA), a PCR variant used to investigate the relatedness between different organisms (117, 118). The second is SSCP (single strand conformational polymorphism), which is employed in mutation studies (119, 120). The same studies can be done by heteroduplex analysis (121, 122). In all cases the running conditions must be carefully controlled. The rate of successful detection of mutations may approach that achieved by temperature gradient gel electrophoresis (123) or DNA sequencing (for a review on DNA sequencing, see 124).

Ethidium bromide, which is usually used for DNA detection, can be added to the running buffer in order to save time. However, the mobilities of DNA-EtBr complexes differ from those of pure DNA molecules and they depend on the mass ratio between

DNA and EtBr (125, unpublished observation). Other intercalating dyes, such as EtBr and thiazole orange dimers, while giving more stable complexes and/or stronger fluorescence, also cause a shift in mobilities (126). The use of these, or other fluorescent tags, such as those linked to 5' end of the primer, allows quantitative detection by dedicated instruments (126). DNA recovery from gels can be done in various ways, as reviewed recently (127).

Precast gels are typically composed of agarose and hydroxyethylated agarose (128), or comprise a new acrylic monomer, N-acryloyl-tris(hydroxymethyl)aminomethane, NAT (129-131). A 12% poly(NAT) gel is displayed in figure 2. Figure 3 shows the DNA pattern in a gel of yet another polymer composition (132). These gels can be run in the presence of ethidium bromide and reused over ten times (unpublished results).

Novel Matrices for Gel Electrophoresis

Acrylamide monomer is a potent neurotoxin with a cumulative effect in man (133), and, with an LD50 of 170 mg/kg in mice (134), it is highly toxic. It is also a carcinogen in mice (133). While polyacrylamide gels are excellent for separating small molecules, DNA fragments above 1000 bp are poorly resolved. Agarose is nontoxic, but the gels

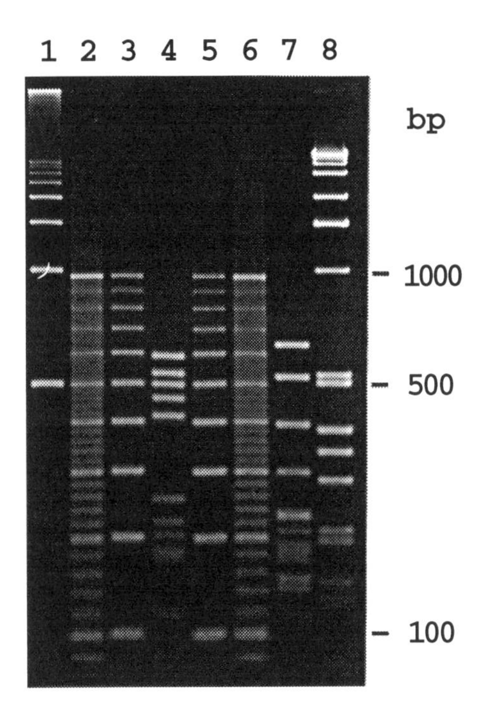

Figure 3. DNA electrophoresis in a gel containing 1% agarose cross-linked with butanediol-diglycidylether (132). The gel was electrophoresed at 7 V/cm for 1 h and 40 min at 20 °C. The samples are 500 bp ladder (Gensura, lane 1), 20 bp ladder (lanes 2 and 6), 100 bp ladder (lanes 3 and 5), pBR322/*Hae* III (lane 4), pBR322/*Msp* I (lane 7), and 1 kb (lane 8). This low percentage gel offers good resolution for short and medium size DNA molecules.

are brittle, they lack optical clarity, cannot be run at elevated temperatures and are unable to resolve well short DNA fragments. These are the major reasons for development of new matrices for gel electrophoresis. It is possible to improve some properties of agarose gels by derivatization and/or partial depolymerization of the agarose polymers (135). Agarose-polyacrylamide composite gels may offer some advantages, but significant improvements can be achieved only with more substantial changes. One such change involves the use of other monomers in place of acrylamide. This approach was pursued by several groups, leading to the discovery of many novel gel formulations based on the existing or newly synthesized acrylic monomers or different polymers (129-131, 136-142). A detailed comparison of the novel matrices is outside the scope of this article.

Future Trends

Our understanding of gel electrophoresis has increased to such a level where the theoretical limitations of the method are becoming apparent. This enables us to fully use the technique's potential and to find new approaches once the limits become unsurmountable. In future, novel matrices, optimized for specific applications, will be widely used. The analysis time will be shortened to seconds through miniaturizations using the microchip technology and new instruments (143). Low cost precast gels will allow gel electrophoretic methods to become standardized and used more frequently for both research and clinical applications.

Acknowledgement

I thank Dr. A.M. Francoeur for helpful suggestions and critical reading of the manuscript.

References

1. Zimm, B.H. and Levene, S.D. 1992. Problems and prospects in the theory of gel electrophoresis of DNA. Q. Rev. Biophys. 25: 171-204.
2. Noolandi, J. 1992. Theory of DNA gel electrophoresis, Adv. Electrophoresis. 5: 1-57.
3. Norden, B., Elvingson, C., Jonsson, M. and Akerman, B. 1991. Microscopic behaviour of DNA during electrophoresis: electrophoretic orientation. Quarterly Rev. Biophys. 24: 103-164.
4. Townsend, K.M. and Dawkins, H.J.S. 1993. Field alternation gel electrophoresis - status quo. J. Chromatogr. 618: 223-249.
5. Lai, E. and Birren, B.W. 1990. Electrophoresis of large DNA molecules: Theory and applications. CSH Press, New York.
6. Rodbard, D. and Chrambach, A. 1970. Unified theory for gel electrophoresis and gel filtration. Proc. Natl. Acad. Sci. USA. 65: 970-977.
7. Ogston, A.G. 1958. The spaces in a uniform random suspension of fibres. Trans. Far. Soc. 54: 1754-1757.

8. Tietz, D. 1988. Evaluation of mobility data obtained from gel electrophoresis: Strategies in the computation of particle and gel properties on the basis of the extended Ogston model. Adv. Electrophoresis. 2: 109-170.
9. Rodbard, D. and Chrambach, A. 1971. Estimation of molecular radius, free mobility, and valence using polyacrylamide gel electrophoresis. Anal. Biochem. 40: 95-134.
10. Lundhal, P., Greijer, E., Sandberg, M., Cardell, S. and Eriksson, K.O. 1986. A model for ionic and hydrophobic interactions and hydrogen-bonding in sodium dodecyl sulfate-protein complexes. Biochim. Biophys. Acta. 873: 20-26.
11. Southern, E.M. 1979. Measurement of DNA length by gel electrophoresis. Anal. Biochem. 100: 319-323
12. Lerman, L.S. and Frisch, H.L. 1982. Why does the electrophoretic mobility of DNA in gels vary with the length of the molecule? Biopolymers. 21: 995-997.
13. Lumpkin, O.J. and Zimm, B.H. 1982. Mobility of DNA in gel electrophoresis. Biopolymers. 21: 2315-2316.
14. Lumpkin, O.J., Dejardin, P. and Zimm, B.H. 1985. Theory of gel electrophoresis of DNA. Biopolymers. 24: 1573-1593.
15. Slater, G.W. and Noolandi, J. 1986. On the reptation theory of gel electrophoresis. Biopolymers. 25: 431-454.
16. DeGennes, P.G. 1971. Reptation of a polymer chain in the presence of fixed obstacles. J. Chem. Phys. 55: 572-579.
17. Hervet, H. and Bean, C.P. 1987. Electrophoretic mobility of lambda phage *Hind*III and *Hae*III DNA fragments in agarose gels: A detailed study. Biopolymers. 26: 727-742.
18. Slater, G.W., Rousseau, J., Noolandi, J., Turmel, C. and Lalande, M. 1988. Quantitative analysis of the three regimes of DNA electrophoresis in agarose gels. Biopolymers. 27: 509-524.
19. Magnusdottir, S., Akerman, B. and Jonsson, M. 1994. DNA electrophoresis in agarose gels: Three regimes of DNA migration identified and characterized by the electrophoretic orientational behavior of DNA. J. Phys. Chem. 98: 2624-2633.
20. Perkins, T.T., Smith, D.E. and Chu, S. 1994. Direct observation of tube-like motion of a single polymer chain. Science. 264: 819-822.
21. Perkins, T.T., Quake, S.R., Smith, D.E. and Chu, S. 1994. Relaxation of a single DNA molecule observed by optical microscopy. Science. 264: 822-826.
22. Schwartz, D.C. and Cantor, C.R. 1984. Separation of yeast chromosome-sized DNAs by pulsed field gradient gel electrophoresis. Cell. 37: 67-75.
23. Deutsch, J.M. 1987. Dynamics of pulsed-field electrophoresis. Phys. Rev. Lett. 59: 1255-1258.
24. Zimm, B.H. 1988. Size fluctuations can explain anomalous mobility in field-inversion electrophoresis of DNA. Phys. Rev. Lett. 61: 2965-2968.
25. Duke, T.A.J. and Viovy, J.L. 1992. Simulation of megabase DNA undergoing gel electrophoresis. Phys. Rev. Lett. 68: 542-545.
26. Burlatsky, S. and Deutch, J. 1993. Influence of solid friction on polymer relaxation in gel electrophoresis. Science. 260: 1782-1784.
27. Duke, T. 1993. Molecular mechanisms of DNA electrophoresis. Int. J. Genome Res. 1: 227-247.
28. Mayer, P., Sturm, J. and Weill, G. 1993. Stretching and overstretching of DNA in pulsed field gel electrophoresis. I. A quantitative study from the steady state birefringence decay. Biopolymers. 33: 1347-1357.

29. Mayer, P., Sturm, J. and Weill, G. 1993. Stretching and overstretching of DNA in pulsed field gel electrophoresis. II. Coupling of orientation and transport in initial response to the field. Biopolymers. 33: 1359-1363.
30. Duke, T. and Viovy, J.L. 1994. Theory of DNA electrophoresis in physical gels and entangled polymer solutions. Physical Review E 49: 2408-2416.
31. Duke, T., Viovy, J.L. and Semenov, A.N. 1994. Electrophoretic mobility of DNA in gels. I. New biased reptation theory including fluctuations. Biopolymers. 34: 239-247.
32. Heller, C., Duke, T. and Viovy, J.L. 1994. Electrophoretic mobility of DNA in gels. II. Systematic experimental study in agarose gels. Biopolymers. 34: 249-259.
33. Schwartz, D.C. and Koval, M. 1989. Conformational dynamics of individual DNA molecules during gel electrophoresis. Nature. 338: 520-522.
34. Bustamante, C., Gurrieri, S. and Smith, S.B. 1990. Observation of single DNA molecules during pulsed-field gel electrophoresis by fluorescence microscopy. Methods. 1: 151-159.
35. Smith, S. B., Gurrieri, S. and Bustamante, C. 1990. Fluorescence microscopy and computer simulations of DNA molecules in conventional and pulsed-filed gel electrophoresis. In: Electrophoresis of Large DNA Molecules: Theory and Applications. E. Lai and B.W. Birren, eds. CSH Laboratory Press. p. 55-79.
36. Gurrieri, S., Rizzarelli, E., Beach, D. and Bustamante, C. 1990. Imaging of kinked configurations of DNA molecules undergoing orthogonal field alternating gel electrophoresis by fluorescence microscopy. Biochemistry. 29: 3396-3401.
37. Smith, S.B., Aldridge, P.K. and Callis, J.B. 1989. Observation of individual DNA molecules undergoing gel electrophoresis. Science. 243: 203-206.
38. Masubuchi, Y., Oana, H., Ono, K., Matsumoto, M., Doi, M., Minagawa, K., Matsuzawa, Y. and Yoshikawa, K. 1993. Periodic behavior of DNA molecules during steady field gel electrophoresis. Macromolecules. 26: 5269-5270.
39. Rampino, N.J. and Chrambach, A. 1991. Conformational correlatives of DNA band compression and bidirectional migration during field inversion gel electrophoresis, detected by quantitative video epifluorescence microscopy. Biopolymers. 31: 1297-1307.
40. Rampino, N.J. 1991. Information concerning the mechanism of electrophoretic DNA separation provided by quantitative video-epifluorescence microscopy. Biopolymers. 31: 1009-1016.
41. Wenisch, E., de Besi, P. and Righetti, P.G. 1993. Conventional isoelectric focusing and immobilized pH gradients in "macroporous" polyacrylamide gels. Electrophoresis. 14: 583-590.
42. Holmes, D.L. and Stellwagen, N.C. 1991. Estimation of polyacrylamide gel pore size from Ferguson plots of normal and anomalously migrating DNA fragments I. Gels containing 3% N,N'-methylenebisacrylamide. Electrophoresis. 12: 253-263.
43. Holmes, D.L. and Stellwagen, N.C. 1991. Estimation of polyacrylamide gel pore size from Ferguson plots of linear DNA fragments II. Comparison of gels with different crosslinker concentrations, added agarose and added linear polyacrylamide. Electrophoresis. 12: 612-619.
44. Stellwagen, N.C. 1987. Electrophoresis of DNA in agarose and polyacrylamide gels. Adv. Electrophoresis. 1: 179-228.

45. Zimm, B.H. and Lumpkin, O. 1993. Reptation of a polymer chain in an irregular matrix; Diffusion and electrophoresis. Macromolecules. 26: 226-234.
46. Zimm, B.H. 1993. Mechanism of gel electrophoresis of DNA: Unexpected findings. Current Opinion in Structural Biol. 3: 373-376.
47. Calladine, C. R., Collis, C. M., Drew, H. R. and Mott, M. R. 1991. A study of electrophoretic mobility of DNA in agarose and polyacrylamide gels. J. Mol. Biol. 221: 981-1005.
48. Bell, L. and Byers, B. 1983. Separation of branched from linear DNA by two-dimensional gel electrophoresis. Anal. Biochem. 130: 527-535.
49. Slater, G.W., Turmel, C., Lalande, M. and Noolandi, J. 1989. DNA gel electrophoresis: Effect of field intensity and agarose concentration on band inversion. Biopolymers 28: 1793-1799.
50. Lalande, M., Noolandi, J., Turmel, C., Brousseau, R., Rousseau, J. and Slater, G.W. 1988. Scrambling of bands in gel electrophoresis of DNA. Nucleic Acids Res. 16: 5427-5437.
51. Doi, M., Kobayashi, T., Makino, Y., Ogawa, M., Slater, G.W. and Noolandi, J. 1988. Band inversion in gel electrophoresis of DNA. Phys. Rev. Lett. 61: 1893-1896.
52. Kozulic, B. 1994. A model of gel electrophoresis. Appl. Theoret. Electrophoresis. In press.
53. Kozulic, B. 1994. On the "door-corridor" model of gel electrophoresis. I. Equations describing the relationship between mobility and size of DNA fragments and protein-SDS complexes. Appl. Theoret. Electrophoresis. In press.
54. Kozulic, B. 1994. On the "door-corridor" model of gel electrophoresis. II. Developments related to new gels, capillary gel electrophoresis and gel chromatography. Appl. Theoret. Electrophoresis. In press.
55. Kozulic, B. 1994. On the "door-corridor" model of gel electrophoresis. III. The gel constant and resistance, and the net charge, friction, diffusion and electrokinetic force of the migrating molecules. Appl. Theoret. Electrophoresis. In press.
56. Bode, H-J. 1976. SDS-polyethyleneglycol electrophoresis: A possible alternative to SDS-polyacrylamide gel electrophoresis. FEBS Lett. 65: 56-58.
57. Bode, H-J. 1977. The use of liquid polyacrylamide in electrophoresis. I. Mixed gels composed of agar-agar and liquid polyacrylamide. Anal. Biochem. 83: 204-210.
58. Bode, H-J. 1977. The use of liquid polyacrylamide in electrophoresis. II. Relationship between gel viscosity and molecular sieving. Anal. Biochem. 83: 364-371.
59. Bode, H-J. 1979. A viscosity model of polyacrylamide gel electrophoresis. Z. Naturforsch. 34c: 512-528.
60. Bode, H-J. 1980. Partitioning and electrophoresis in flexible polymer networks. In: Electrophoresis '79. B.J. Radola, ed. W. de Gruyter, Berlin. p 39-52.
61. Chu, G. 1991. Bag model for DNA migration during pulsed-field electrophoresis. Proc. Natl. Acad. Sci. USA. 88: 11071-11075.
62. Luckey, J.A. and Smith, L.M. 1993. A model for the mobility of single-stranded DNA in capillary gel electrophoresis. Electrophoresis. 14: 492-501.
63. Figeys, D. and Dovichi, N.J. 1993. Mobility of single-stranded DNA as a function of cross-linker concentration in polyacrylamide capillary gel electrophoresis. J. Chromatogr. 645: 311-317.

64. Ganzler, K., Greve, K.S., Cohen, A.S., Karger, B.L., Guttman, A. and Cooke, N.C. 1992. High-performance capillary electrophoresis of SDS-protein complexes using UV-transparent polymer networks. Anal. Chem. 64: 2665-2671.
65. Pariat, Y.F., Berka, J., Heiger, D.N., Schmitt, T., Vilenchik, M., Cohen, A.S., Foret, F. and Karger, B.L. 1993. Separation of DNA fragments by capillary electrophoresis using replaceable linear polyacrylamide matrices. J. Chromatogr. 652: 57-66.
66. Ruiz-Martinez, M.C., Berka, J., Belenkii, A., Foret, F., Miller, A.W. and Karger, B.L. 1993. DNA sequencing by capillary electrophoresis with replaceable linear polyacrylamide and laser-induced fluorescence detection. Anal. Chem. 65: 2851-2858.
67. Grossman, P.D. 1994. Electrophoretic separation of DNA sequencing extension products using low-viscosity entangled polymer networks. J. Chromatogr. A 663: 219-227.
68. Bae, Y.C. and Soane, D. 1993. Polymeric separation media for electrophoresis: Cross-linked systems or entangled solutions. J. Chromatogr. 652: 17-22.
69. Barron, A.E., Blanch, H.W. and Soane, D.S. 1994. A transient entanglement coupling mechanism for DNA separation by capillary electrophoresis in ultradilute polymer solutions. Electrophoresis. 15: 597-615.
70. Johnson, P. H., Miller, M. J. and Grossman, L. I. 1980. Electrophoresis of DNA in agarose gels. II. Effects of loading mass and electroendosmosis on electrophoretic mobilities. Anal. Biochem. 102: 159-162.
71. Meyers, J. A., Sanchez, D., Elwell, L. P. and Falkow, S. 1976. Simple agarose gel electrophoretic method for the identification and characterization of plasmid deoxyribonucleic acid. J. Bacteriol. 127: 1529-1537.
72. McNally, L., Baird, M., McElfresh, K., Eisenberg, A. and Balazs, I. 1990. Increased migration rate observed in DNA from evidentiary material precludes the use of sample mixing to resolve forensic cases of identity. Appl. Theoret. Electrophoresis. 1: 267-272.
73. Stellwagen, N.C. and Stellwagen, J. 1989. Orientation of DNA and the agarose gel matrix in pulsed electric fields. Electrophoresis. 10: 332-344.
74. Holmes, D.L. and Stellwagen, N.C. 1989. Electrophoresis of DNA in oriented agarose gels. J. Biomol. Structure and Dynamics. 7: 311-327.
75. Stellwagen, N.C. and Stellwagen, J. 1993. "Flip-flop" orientation of agarose gel fibers in pulsed alternating electric fields. Electrophoresis. 14: 355-368.
76. Stellwagen, J. and Stellwagen, N.C. 1994. Transient electric birefringence of agarose gels. I. Unidirectional electric fields. Biopolymers. 34: 187-201.
77. Mayer, P., Slater, G.W. and Drouin, G. 1993. Exact behaviour of single-stranded DNA electrophoretic mobilities in polyacrylamide gels, Appl. Theoret. Electrophoresis. 3: 147-155.
78. Doggett, N.A., Smith, C.L. and Cantor, C.R. 1992. The effect of DNA concentration on mobility in pulsed field gel electrophoresis. Nucl. Acids Res. 20: 859-864.
79. Smith, S. S., Gilroy, T. E. and Ferrari, F. A. 1983. The influence of agarose-DNA affinity on the electrophoretic separation of DNA fragments in agarose gels. Anal. Biochem. 128: 138-151.
80. Upcroft, P. and Upcroft, J.A. 1993. Comparison of properties of agarose for electrophoresis of DNA. J. Chromatogr. 618: 79-93.
81. Wicar, S., Vilenchik, M., Belenkii, A., Cohen, A.S. and Karger, B.L. 1992. Influence of coiling on performance in capillary electrophoresis using open tubular and polymer network columns. J. Microcol. Sep. 4: 339-348.

82. Righetti, P.G. 1983. Isoelectric Focusing: Theory, Methodology and Applications. Elsevier, Amsterdam.
83. Righetti, P.G. 1990. Immobilized pH Gradients: Theory and Methodology. Elsevier, Amsterdam.
84. Hames, B.D., and Rickwood, D. 1990. Gel Electrophoresis of Proteins. A Practical Approach, 2nd ed. IRL Press, Oxford.
85. Westermeier, R. 1993. Electrophoresis in Practice. VCH, Weinheim.
86. Dunbar, B.S. 1988. Two-Dimensional Electrophoresis and Immunological Techniques. Plenum Press, New York.
87. Lambin, P. and Fine, J.M. 1979. Molecular weight estimation of proteins by electrophoresis in linear polyacrylamide gradient gels in the absence of denaturing agents. Anal. Biochem. 98: 160-168.
88. Rothe, G.M. 1991. Determination of the size, isomeric nature and net charge of enzymes by pore gradient gel electrophoresis. Adv. Electrophoresis 4: 251-358.
89. Kozulic, B., Kappeli, O., Meussdoerffer, F. and Fiechter, A. 1987. Characterization of a soluble carnitine acetyltransferase from *Candida tropicalis*. Eur. J. Biochem. 168: 245-250.
90. Kozulic, B., Mosbach, K. and Meussdoerffer, F. 1988. Biosynthesis of soluble carnitine acetyltransferases from the yeast *Candida tropicalis*. Biochem. J. 253: 845-849.
91. Kozulic, B., Leustek, I., Pavlovic, B., Mildner, P. and Barbaric, S. 1987. Preparation of the stabilized glycoenzymes by cross-linking their carbohydrate chains. Appl. Biochem. Biotechnol. 15: 265-278.
92. Heimgartner, U., Kozulic, B. and Mosbach, K. 1990. Reversible and irreversible cross-linking of immunoglobulin heavy chains through their carbohydrate residues. Biochem. J. 267: 585-591.
93. Schagger, H., Cramer, W.A. and von Jagow, G. 1994. Analysis of molecular masses and oligomeric states of protein complexes by blue native electrophoresis and isolation of membrane protein complexes by two-dimensional native electrophoresis. Anal. Biochem. 217: 220-230.
94. Lambin, P. 1978. Reliability of molecular weight determination of proteins by polyacrylamide gradient gel electrophoresis in the presence of sodium dodecyl sulfate. Anal. Biochem. 85: 114-125.
95. Hames, B.D. and Rickwood, D. 1981. Gel Electrophoresis of Proteins. A Practical Approach. IRL Press, Oxford. p. 25.
96. Kumar, T.K.S., Gopalakrishna, K., Prasad, V.V.H. and Pandit, M.W. 1993. Multiple bands on the sodium dodecyl sulfate - polyacrylamide gel electrophoresis gels of proteins due to intermolecular disulfide cross-linking, Anal. Biochem. 213: 226-228.
97. Chiari, M., Micheletti, C., Righetti, P.G. and Poli, G. 1992. Polyacrylamide gel polymerization under non-oxidizing conditions, as monitored by capillary zone electrophoresis. J. Chromatogr. 598: 287-297.
98. Singh, R. 1994. Odorless SDS-PAGE of proteins using sodium 2-mercaptoethanesulfonate. BioTechniques. 17: 263-265.
99. Fernandez-Patron, C., Castellanos-Serra, L. and Rodriguez, P. 1992. Reverse staining of sodium dodecyl sulfate polyacrylamide gels by imidazole-zinc salts: Sensitive detection of unmodified proteins, BioTechniques. 12: 564-573.

100. Gabriel, O. and Gersten, D.M. 1992. Staining for enzymatic activity after gel electrophoresis. I. Anal. Biochem. 203: 1-21.
101. Pepinsky, R.B. 1983. Localization of lipid-protein and protein-protein interactions within the murine retrovirus *gag* precursor by a novel peptide-mapping technique. J. Biol. Chem. 258: 11229-11235.
102. Saris, C.J.M., Van Eenbergen, J., Jenks, B.G. and Bloemers, H.P.J. 1983. Hydroxylamine cleavage of proteins in polyacrylamide gels. Anal. Biochem. 132: 54-67.
103. Leisola, M.S.A., Kozulic, B., Meussdoerffer, F. and Fiechter, A. 1987. Homology among multiple extracellular peroxidases from *Phanerochaete chrysosporium.* J. Biol. Chem. 262: 419-424.
104. Bhavsar, J.H., Remmler, J. and Lobel, P. 1994. A method to increase efficiency and minimize anomalous electrophoretic transfer in protein blotting. Anal. Biochem. 221: 234-242.
105. Beck, S. 1988. Protein blotting with direct blotting electrophoresis. Anal. Biochem. 170: 361-366.
106. Collier, C.F., Regester, D.J. and Robertson, C.W. 1993. Apparatus for direct blotting and automated electrophoresis, transfer and detection and processes utilizing the apparatus thereof. US Patent 5,234,559.
107. Patterson, S.D. 1994. From electrophoretically separated protein to identification: Strategies for sequence and mass analysis. Anal. Biochem. 221: 1-15.
108. Kozulic, B. and Heimgartner, U. 1991. An apparatus for submerged gel electrophoresis, Anal. Biochem. 198: 256-262.
109. Kozulic, B. and Heimgartner, U. 1993. Apparatus and method for submerged gel electrophoresis. US Patent 5,259,943.
110. Kreisher, J.H., Belle Isle, H.D. and Nalbantian, C.A. 1986. Horizontal gel electrophoresis device. US Patent 4,588,491.
111. Audeh, Z.L. 1987. Non-mechanical buffer circulation apparatus for electrophoresis. US Patent 4,702,814.
112. Day, I.N.M. and Humphries, S.E. 1994. Electrophoresis for genotyping: Microtiter array diagonal gel electrophoresis on horizontal polyacrylamide gels, Hydrolink, or agarose. Anal. Biochem. 222: 389-395.
113. Hagerman, P.J. 1990. Sequence-directed curvature of DNA. Annu. Rev. Biochem. 59: 755-781.
114. Diekmann, S. 1989. The migration anomaly of DNA fragments in polyacrylamide gels allows the detection of small sequence-specific DNA structure variations. Electrophoresis. 10: 354-359.
115. Kozulic, B. 1994. Looking at bands from another side. Anal. Biochem. 216: 253-261.
116. Anonymous. 1994. Resolutions, Volume 10 (4). FMC Corporation, Rockland, ME, USA. p4.
117. Bostock, A., Khattak, M.N., Matthews, R. and Burne, J. 1993. Comparison of PCR fingerprinting, by random amplification of polymorphic DNA, with other molecular typing methods for *Candida albicans*. J. Gen. Microbiol. 139: 2179-2184.
118. Wang, G., Whittam, T.S., Berg, C.M. and Berg, D.E. 1993. RAPD (arbitrary primer) PCR is more sensitive than multilocus enzyme electrophoresis for distinguishing related bacterial strains. Nucl. Acids Res. 21: 5930-5933.

119. Orita, M., Iwahana, H., Kanazawa, H., Hayashi, K. and Sekiya, T. 1989. Detection of polymorphisms of human DNA by gel electrophoresis as single-strand conformation polymorphism. Proc. Natl. Acad. Sci. USA. 86: 2766-2770.
120. Vidal-Puig, A.and Moller, D.E. 1994. Comparative sensitivity of alternative single-strand conformation polymorphism (SSCP) methods. BioTechniques. 17: 490-496.
121. Bidwell, J., Wood, N., Clay, T., Pursall, M., Culpan, D., Evans, J., Bradley, B., Tyfield, L., Standen, G. and Hui, K. 1994. DNA heteroduplex technology. Adv. Electrophoresis. 7: 311-351.
122. Ganguly, A., Rock, M.J. and Prockop, D.J. 1993. Conformation-sensitive gel electrophoresis for rapid detection of single-base differences in double-stranded PCR products and DNA fragments: Evidence for solvent-induced bends in DNA heteroduplexes. Proc. Natl. Acad. Sci. USA. 90: 10325-10329.
123. Riesner, D., Henco, K. and Steger, G. 1991. Temperature-gradient gel electrophoresis: A method for the analysis of conformational transitions and mutations in nucleic acids and proteins. Adv. Electrophoresis. 4: 171-250.
124. Griffin, H.G. and Griffin, A.M. 1993. DNA sequencing protocols. Humana Press, Totowa, NJ.
125. Waye, J.S.and Fourney, R.M. 1990. Agarose gel electrophoresis of linear genomic DNA in the presence of ethidium bromide: Band shifting and implications for forensic identity testing. Appl. Theoret. Electrophoresis. 1: 193-196.
126. Rye, H.S., Yue, S., Quesada, M.A., Haugland, R.P., Mathies, R.A. and Glazer, A.N. 1993. Picogram detection of stable dye-DNA intercalation complexes with two-color laser-excited confocal fluorescence gel scanner. Methods Enzymol. 217: 414-431.
127. Duro, G., Izzo, V. and Barbieri, R. 1993. Methods for recovering nucleic acid fragments from agarose gels. J. Chromatogr. 618: 95-104.
128. Guiseley, K.B. 1976. Modified agarose, and agar and method of making same. US Patent 3,956,273.
129. Kozulic, M., Kozulic, B. and Mosbach, K. 1987. Poly-N-acryloyl-Tris gels as anticonvection media for electrophoresis and isoelectric focusing. Anal. Biochem. 163: 506-512.
130. Kozulic, B., Mosbach, K. and Pietrzak, M. 1988. Electrophoresis of DNA restriction fragments in poly-N-acryloyl-Tris gels. Anal. Biochem. 170: 478-484.
131. Kozulic, B. and Mosbach, K. 1994. Polymers, and their use as gels for electrophoresis. US Patent 5,319,046.
132. Kozulic, B. 1994. Cross-linked linear polysaccharide polymers as gels for electrophoresis. European Patent Application 0 604 862 A2.
133. Bailey, E., Farmer, P.B., Bird, I., Lamb, J.H. and Peal, J.A. 1986. Monitoring exposure to acrylamide by the determination of S-(2-carboxyethyl)cysteine in hydrolyzed hemoblobin by gas chromatography-mass spectrometry. Anal. Biochem., 157: 241-248.
134. Windholz, M. 1983. The Merck Index, 10-th edition. Merck & Co. Inc., Rahway, NJ, USA. p. 19.
135. Nochumson, S., Curtis, F.P., Morgan, J.H. and Kirkpatrick, F.H. 1992. Polysaccharide resolving gels and gel systems for stacking electrophoresis. US Patent 5,143,646.

136. Kozulic, B. 1993. Hydrophilic synthetic gels and their use in electrophoresis. US Patent 5,202,007.
137. Kozulic, B. and Heimgartner, U. 1993. Hydrophilic and amphiphatic monomers, their polymers and gels and hydrophobic electrophoresis. US Patent 5,185,466.
138. Shorr, R. 1993. Electrophoretic media. US Patent 5,219,923.
139. Molinari, R.J., Connors, M. and Shorr, R.G.L. (1993) Hydrolink gels for electrophoresis. Adv. Electrophoresis 6: 43-60.
140. Righetti, P.G. 1993. New formulations for polyacrylamide matrices in electrophoretic and chromatographic methodologies. International Patent Application WO 93/11174.
141. Zewert, T. and Harrington, M. 1994. Acrylic polymer electrophoresis support media. US Patent 5,290,411.
142. Ponticello, I.S. and LaTart, D.B. 1993. Electrophoresis element comprising a polymer containing a haloacetamido group. US Patent 5,212,253.
143. Manz, A., Verpoorte, E., Effenhauser, C.S., Burggraf, N., Raymond, D.E. and Widmer, H.M. 1994. Planar chip technology for capillary electrophoresis. Fresenius J. Anal. Chem. 348: 567-571.

From: *Molecular Biology: Current Innovations and Future Trends.*
ISBN 1-898486-01-8 ©1995 Horizon Scientific Press, Wymondham, U.K.

5

PULSED FIELD GEL ELECTROPHORESIS

Alexander Kolchinsky and Roel Funke

Abstract

Pulsed field gel electrophoresis extends the size range of resolution of DNA molecules to many megabases. This bridges a gap that existed in techniques for the structural analysis of large genomes. The ability to analyze large molecules of DNA has enabled the development of cloning vectors with large capacities such as yeast artificial chromosomes, and permitted applications including the transfer of artificial chromosomes containing intact gene clusters to mammalian cells. It has also brought the creation of artificial chromosomes for mammals and plants into the realm of feasibility.

Introduction

Pulsed field gel electrophoresis (PFGE) resolves molecules of DNA in the range of 10 kilobases (kb) to several megabases (Mb). This cannot be achieved by conventional electrophoretic techniques. PFGE was developed twelve years ago by Schwartz *et al.* (1). Since then, both protocols and equipment have become routine in almost every laboratory working with DNA, and a comprehensive practical manual has been published (2). This book covers virtually every aspect of PFGE, contains 900 references, and is written in an extremely user-friendly manner. Another useful manual was published by Burmeister and Ulanovsky (3). Therefore, we will discuss only very recent publications in light of the most important applications of the technique and will present a protocol based on our experience that can be easily accommodated for other tissues and purposes.

DNA molecules separated in agarose under a constant electric field are retarded by the agarose matrix. This retention correlates inversely with the size of the molecule: DNA molecules adopt a loose compact conformation, and the longer the molecule, the larger the globule. However, longer DNA molecules adopt a stable rod-like conformation and their cross-section and retention by the agarose matrix are not determined by their size. As a result, all DNA molecules above a particular cut-off size have the same mobility. The cut-off size is determined by the concentration of agarose and conditions of electrophoresis, but rarely exceeds 50 kb. In PFGE, DNA is subjected to an electric field that periodically changes its direction. With every change, the random coils of DNA molecules change their conformation and orientation. This process is called relaxation. The time needed by individual molecules to realign themselves with the new field before recommencing their movement in that direction depends on their size. The molecules follow a zigzag path, and their net movement down the gel is again inversely correlated with their size. DNA resolved by PFGE has a limit of resolution which is determined primarily by the pulse time, and to a lesser extent by other parameters of the run. All fragments above a specific size move together forming the so-called compression zone (CZ), or zone of limiting resolution. For instance, at a 2 sec pulse time in 0.25 x TBE the CZ comprises fragments longer than approximately 70 kb; at 10 sec, 300 kb; at 1 min, 1 Mb; at 10 min, 3 Mb.

There are many designs of gel boxes for PFGE based on different geometries of electric fields (2). The simplest of these is the field-inversion gel electrophoresis (FIGE) and the most popular is the contour-clamped homogenous gel electrophoresis (CHEF). FIGE can be performed in any conventional horizontal chamber. The electric field is periodically inverted and forward movement of DNA is achieved either by longer pulses or higher voltage applied in one direction. This system can be built in any electric shop by inserting a switching device between the power pack and the chamber. Since the angle of the electric field in FIGE does not change, the lanes are straight and parallel. Its disadvantages are that it gives somewhat fuzzy bands above 500-750 kb and generates artifactual inversion of mobility for some size ranges, with fragments moving slower than expected. CHEF exploits hexagonal geometry of electrodes (Fig. 1), and the lanes are straight and parallel (see below). In the manufactured device CHEF-DRIII made by BioRad Laboratories, the voltage applied to each of 24 electrodes is adjusted in order to provide even better uniformity across the gel. The angle between the fields is maintained at 120° in an older device, CHEF-DRII, and is adjustable in newer systems CHEF-DRIII and CHEF-MAPPER to optimize separation of very large fragments.

It was mentioned that the range of resolution of PFGE is determined primarily by pulse time. To improve the linearity of resolution, a gradually changing (ramped) pulse time throughout the run is often used. This also solves the problem of size reversal that is sometimes encountered in pulsed field migration.

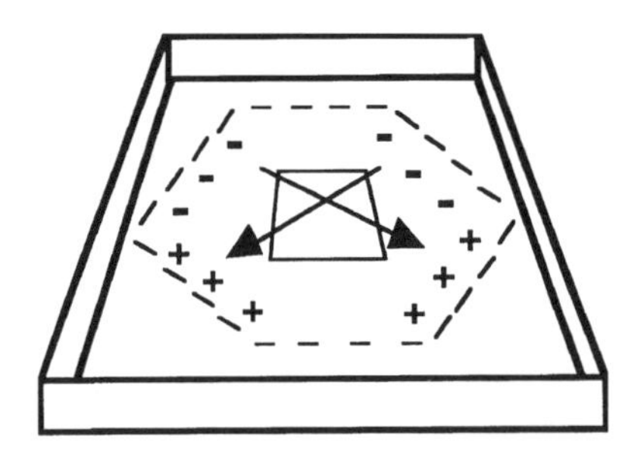

Figure 1. Scheme of the gelbox for CHEF. Electrodes are located in a hexagonal configuration, the gel is shown as a rectangle in the middle. Alternating electric fields are shown by crossed arrows.

PFGE: Current Applications

PFGE became an indispensable means to bridge the gap between three levels of structural studies of eukaryotic genomes: genetic, cytogenetic, and molecular. Optical resolution of metaphase chromosomes usually corresponds to several megabases of

DNA, while conventional electrophoresis resolves only fragments up to 50 kb. The resolution of genetic analysis in higher eukaryotes rarely goes beyond 1-2 cM (centiMorgan, the elementary unit of genetic recombination), which again corresponds to megabase(s) of DNA. Thus, PFGE enables researchers to construct physical maps with resolutions spanning these extremes.

A technique that was enabled by PFGE and became a major tool in genome research was cloning of DNA in yeast artificial chromosomes [YACs; for review see (4)]. Prior to the introduction of YACs the largest fragments amenable to cloning did not exceed 40 kb; YACs can hold 100 kb-2 Mb of foreign DNA. Naturally, the analysis of these cloned inserts was unthinkable without PFGE. Later, cloning in YACs was supplemented with alternative procedures for cloning large DNA fragments (4). These cloning techniques covered the gap between the emerging power of physical mapping and limited size of cloned DNA fragments.

Integration of genetic and physical maps enables one to undertake positional cloning of a gene of interest. The prerequisites for positional cloning include a well defined gene that confers the phenotype of interest, a set of molecular markers tightly linked to the gene, a physical map of the corresponding segment with molecular markers on the map, and a library of large DNA inserts cloned, for instance, in YACs. Because of the resolution range of PFGE, a detailed genetic map around the gene can be correlated with the physical map, and the genomic region encompassing the gene can be cloned in YACs.

With this in view, it is quite important to determine the correlation between physical and genetic distances in the genome to estimate the approximate number of linked molecular markers required for complete coverage, and length of DNA to be cloned in search of a gene starting from linked molecular markers. A preliminary estimate can be produced simply by dividing the total physical size of a haploid genome by its genetic size deduced from its detailed genetic map, if available. However, real mapping shows that this value varies enormously throughout the genome. First, it was found recently that telomeric segments of chromosomes are highly enriched in genes (5). The term gene is used here in its traditional meaning, an elementary unit of recombination. In other words, recombination frequency near the ends of chromosomes is increased. In addition, fluctuations of recombination rate along chromosomes were estimated for many organisms. For example, physical and genetic mapping of a 470 kb segment of maize DNA containing two genes showed that recombination frequencies in the immediate vicinities of the genes are increased as compared with intergenic intervals (6). The distances are 217 kb/cM for a 1 kb segment inside the locus, and 1560 kb/cM for the intergenic region.

Physical mapping in chromosome areas with the lowest recombination frequencies, near centromeres, is hampered by a shortage of DNA markers in those regions. Centromeric areas consist of arrays of tandem repeats with little variability. In some cases molecular markers for these areas have to be found by the analysis of large fragments of DNA resolved by PFGE (7); for review see also reference 8.

In some cases physical mapping goes to the margin of PFGE resolution. Although resolution of fungal chromosomes up to 10-12 Mb long has been reported, it is difficult to resolve fragments longer than 2-2.5 Mb obtained by restriction digestion and revealed by hybridization to molecular probes. The longest restriction fragments used in mapping, that we were able to find, reached 5.7 Mb (9). The electrophoresis took six days because at this resolution, the applied voltage gradient must be very low. The authors were able to locate an important gene for ataxia-telangiectasia on a 3 Mb fragment between two molecular markers.

Physical mapping of large genomes encounters several major problems. First, ordering of sites for the same restriction enzyme requires partial digestion of DNA. In many cases, this is achieved naturally, because of incomplete and/or variable methylation of the restriction sites. Most rare cutting restriction enzymes are sensitive to methylation of the sequences CpG found in animals and CpG and CpXpG found in plants. In addition, different sources of high molecular weight DNA used for mapping, like cell lines in mammals and varieties in plants, may show variable methylation patterns. We found marked differences even in different batches of soybean leaves, probably because of minor variations in growth conditions (unpublished). This complicates the interpretation of the data, however it can be very useful in the analysis of partial digests, for example in the detailed mapping of kallikrein genes in rat genome (10).

A more difficult task is to establish colinearity of cloned products of partial digests with their genomic counterpart. This can be achieved by so-called RARE mapping (11). First, DNA around the *Eco*RI sites that are to be mapped is sequenced and a complementary oligonucleotide is synthesized. In the presence of the RecA protein, it replaces one strand in the target site and protects it from methylation, while the rest of the DNA is methylated with *Eco*RI-methylase. Then the oligonucleotide and proteins are washed away and the DNA is treated with *Eco*RI, which is expected to introduce just one cut at the non-methylated site. By this approach Ferrin and Camerini-Otero (12) and Negorev *et al.* (13) were able to establish colinearity of cloned fragments to their sources.

Often markers from molecular maps detect a mixture of monomorphic and polymorphic bands in restriction digests used to score polymorphisms. When the same marker shows multiple bands in rare-cutter restriction digests, it is difficult to determine which band corresponds to the polymorphic copy of the marker and whether some of the bands represent partial digestion products. By digesting the entire lane of PFGE resolved DNA with the enzyme that revealed the polymorphic band and separating the products on a conventional gel perpendicular to the direction of the first electrophoresis, it is possible to assign copies of the marker to individual bands on the PFGE Southern. This innovation was originally used to produce a physical map of the immunoglobulin heavy chain variable region genes of the human genome (29), and has been adapted for genomic analysis of tandemly repeated DNA (7) and physical mapping of homologous chromosomal regions of the soybean genome (14)

Partial digests become more important in the mapping of YAC inserts. There is a growing consensus that so-called CpG islands are the best landmarks for genes in eukaryotic DNA (15). The mapping procedure can be illustrated by work recently published by Coleman *et al.* (16). After finding YACs in the area of interest, the authors generated partial digests with restriction enzymes that are enriched for CpG sequences in their recognition sites, such as *Not*I, *Bss*HII, *Eag*I and *Sac*I. CpG islands show coincidence of at least three different restriction sites within a very short distance. The result looks like a band of partial products that spans several lanes on the Southern blot.

Occasionally, PFGE fragments are used as molecular markers. It applies to simple tandem repeats that cannot be converted into shorter informative hybridizing fragments. One application of these markers to map centromeres was mentioned above, it also proved very useful in the mapping of plant telomeres (8). Plant telomeres consist of

7 bp telomeric repeats that form arrays up to 60 kb long and subtelomeric tandem satellite repeats that can occupy up to several hundred kb. Both of these tandem repeats were successfully used in several plant species to generate molecular markers and put them on existing maps (8).

Sometimes PFGE is used to produce electrophoretic karyotypes of DNA from organisms with small genomes. Single chromosomes of bacteria or eukaryotic organelles are restricted by a rare cutting enzyme and the resulting fragments resolved by PFGE. Use of partial digests and/or several enzymes makes it feasible to construct a physical map of the molecule of interest. The resolution of individual chromosomes of eukaryotes is usually more challenging because of their size. For example, authors of a recent paper were able to resolve all eight chromosomes of *Aspergillus niger* measuring up to 6.6 Mb (17). Because some of the chromosomes have the same mobility on PFGE, the authors constructed strains in which they artificially enlarged individual chromosomes by integrating multiple copies of the *glaA* gene. As a result they were able to differentiate the chromosomes by PFGE and accurately align the genetic and physical maps by Southern hybridization. One PFGE run took six days, as in a physical mapping case mentioned above.

Although the great majority of physical mapping by PFGE uses hybridization to reveal fragments complementary to the probe, there is an alternative approach based on PCR. Pieces of pulsed-field gels in the desired size range can be used in PCR reactions directly, after melting and dilution of agarose to prevent solidification, or after treatment with agarase. These fractions can be amplified with arbitrary primers for further cloning (18) or for localization of a fragment corresponding to the probe of interest (19).

Future Trends

Recently, it became possible to transfer large segments of DNA cloned in YACs into mammalian cells. This is important for several reasons. First, the expression of mammalian genes in transgenic cells depends on flanking sequences (20, 21), and incorporation of a large segment surrounding the gene of interest alleviates these fluctuations. Second, some mammalian genes are so large that functional copies can be cloned and handled in YACs or other vectors of similar capacity only (22). Third, some important genes form large clusters and must be transferred together for proper functioning and further studies, for example transfer of HLA genes (23). Eventually, the transfer of large segments of DNA and assembly of functional artificial chromosomes of higher plants and animals will be one of the most challenging applications of PFGE in genetic research.

The creation of vectors for gene therapy will make use of PFGE technology. In recent work, a combination of an adenovirus and a YAC was obtained that could prove a useful vector for genetic manipulations in yeast cells and for the delivery of genes into mammalian cells for gene therapy (24).

Protocols

The following protocols include all techniques necessary for the preparation of DNA for physical mapping or cloning and two-dimensional electrophoresis.

Preparation of Megabase DNA in Plugs

DNA molecules in solution are extremely sensitive to shearing, and isolation procedures that involve pipetting and ethanol precipitation usually yield DNA with an average molecular weight 20-50 kb. To avoid handling DNA in liquid, suspensions of live cells are mixed with low melting point agarose and embedded in plugs or blocks. The embedded cells are then lysed with detergent in a high-EDTA buffer. Nucleases and cellular debris are removed by diffusion in the presence of proteinase K, leaving high molecular weight chromosomal DNA essentially intact and accessible to restriction enzymes.

This procedure was first described by Schwartz and Cantor (25), who used it to produce the first electrophoretic karyotype of yeast. It has been adapted to study DNA from many organisms. Plants pose a special problem because of the cell wall which must be removed. High nuclease activities result in a certain amount of degradation of the DNA, and deposits of starch or phenolic compounds in the agarose blocks may interfere with the diffusion or activity of restriction enzymes. Procedures have been developed for the isolation of megabase DNA from nuclei. This is reported to be an effective method with the added advantage that chloroplast and mitochondrial DNA are not included in the DNA plugs or microbeads (26, 27). The following protocol was designed for the preparation of megabase DNA starting from soybean leaf tissue, but may be readily adapted for other organisms. The density at which the cells are embedded should be adjusted depending on the nuclear DNA content of the experimental organism and on the desired final DNA concentration in the plugs. For accurate sizing of high molecular weight fragments, DNA concentration should not be too high (80-120 μg/ml). When DNA concentration is 200 μg/ml or more, the apparent size of fragments is significantly greater than their true size (28).

1. Young expanding trifoliate leaves (1 cm or less in length) or primary leaves are harvested for the preparation of protoplasts. Healthy appearance of plants is extremely important for successful isolation. Surface sterilized 1-2 mm leaf strips are digested with 2% driselase (Sigma Chemicals), 0.2% pectolyase Y-10 (Sigma Chemicals), and 1% cellulase Onozuka R-10 (Serva) for 5 hours with gentle shaking in a buffer (PB) containing 0.4 M mannitol, 10 mM $CaCl_2$, 5 mM MES, and 5 mM ascorbic acid, pH 5.6.
2. Following digestion the mixture is filtered through a 43 mm nylon mesh and protoplasts are collected by centrifugation at 100g for 10 min. The cells are rinsed once with PB. Cells are counted and after a second centrifugation step gently resupended in PB. The suspension is then mixed with 1.5% agarose in PB kept at 45 °C, and immediately cast into a chilled mould. Moulds can be designed to suit individual preferences, and our DNA plugs have a volume of 150-200 μl depending on the tooth size of the comb we plan to use. The final concentration of cells should be 2 to 3 x 10^7 per ml. Assuming that the diploid nucleus of soybean contains about 5 pg of DNA, this

yields plugs with a nuclear DNA concentration of about 100 to 150 μg/ml. The total DNA concentration in the plugs is somewhat higher if the plastid and mitochondrial genomes are taken into consideration.
3. Plugs are transferred to three times their combined volume of ESP (0.5M EDTA, 1% sarkosyl, 1 mg/ml proteinase K) at 50 °C while being gently agitated at 50 rpm. After 24 hours the solution is replaced with fresh ESP and left for a further 24 hours, at which point the plugs can be prepared for restriction digestion.

Restriction Digestion and Electrophoresis

1. Proteinase K is inactivated by washing the plugs twice for 30 min in 10-20 volumes of TE buffer containing 1 mM PMSF (100 μl of 100 mM freshly prepared acetone stock per 10 ml of TE) at 50 °C. The plugs are rinsed three more times with fresh changes of TE without PMSF.
2. One-third to one-quarter of a 200 μl plug (about 5-10 μg of DNA) is weighed in a tared microcentrifuge tube. Treating the plug as void volume, water and 10x restriction buffer are added to make 100 μl of 1x buffer and the tube is placed on ice for 30 min. The liquid is removed and replaced with 50 μl 1x buffer and 10-20 units restriction enzyme per μg DNA. The mixture is left at 4 °C for 30 min to allow the enzyme to diffuse into the plugs before becoming fully active. Digestions are carried out at the recommended temperature for at least 6 hours, and terminated with the addition of 1 ml of ice cold TE. Plugs are sealed into the wells of a gel with a small amount of melted agarose.
3. Conditions for pulsed field gel resolution will depend on the size range of fragments produced by the restriction enzymes and the desired range of resolution. Some examples of restriction digests of embedded soybean chromosomal DNA resolved by PFGE are shown in Figure 2. Switch time, temperature, buffer concentration, agarose concentration, and salt concentration in the sample block are among the factors that influence the mobility and resolution of DNA fragments in pulsed field gel electrophoresis (2, 31).
4. It is important to include on the gel size standards appropriate to the range of electrophoretic resolution. In the range 200 kb to 2.2 Mb, it is convenient to use chromosomes of *Saccharomyces cerevisiae;* phage lambda concatemers are suitable size standards for the range 50 - 1000 kb. These may be purchased or prepared. We prefer standards from Sigma (St. Louis, MO), which come in convenient syringes. Before preparing yeast standards from cells, one should make sure that the size of chromosomes of the strain in use is well characterized.

Hybridization

The conditions for transfer and hybridization of pulsed field gels are similar to those for conventional gels, with the provision that high molecular weight DNA fragments can be nicked efficiently and consistently. Different laboratories use combinations and modifications of procedures that involve UV irradiation and treatment with 0.25 M HCl. The protocol we supply below works well in our hands, but conditions should be optimized for each laboratory. To do this, it is convenient to use lanes of resolved yeast chromosomes from a single gel. The DNA in each lane is nicked using different procedures (different lengths of time of HCl treatment, dose of UV irradiation, or combinations thereof), the gel is put back together and transferred as a whole. The whole gel is probed with labeled yeast chromosomal DNA and the signal quality obtained with each treatment can be directly compared.

1. Stain the gel for 30 min with 1 μg/ml ethidium bromide. It is important for ultraviolet cleavage of large DNA that all the fragments throughout the whole thickness of the gel are stained, since only DNA bound by ethidium bromide will be nicked. For clear photographs, destain the gel for at least one hour.
2. The DNA is nicked with 1500 μJ UV in a UV Stratalinker™ 1800, followed by depurination with 0.25M HCl for 20 min.
3. After briefly rinsing the gel with distilled water, the DNA is denatured for 30 min with 0.4 M NaOH.
4. The DNA is vacuum-transferred to membranes (BioTrans, ICN; or Zeta-Probe, BioRad) in 10 x SSC for 2 h at 40 cm H_2O (Vacugene, LKB).
5. Following transfer the membranes are rinsed in 2 x SSC and the DNA is crosslinked to the membrane by irradiation with 1200 μJ UV light. Figure 2 shows an example of hybridization of a PFGE Southern blot probed with a soybean molecular marker.

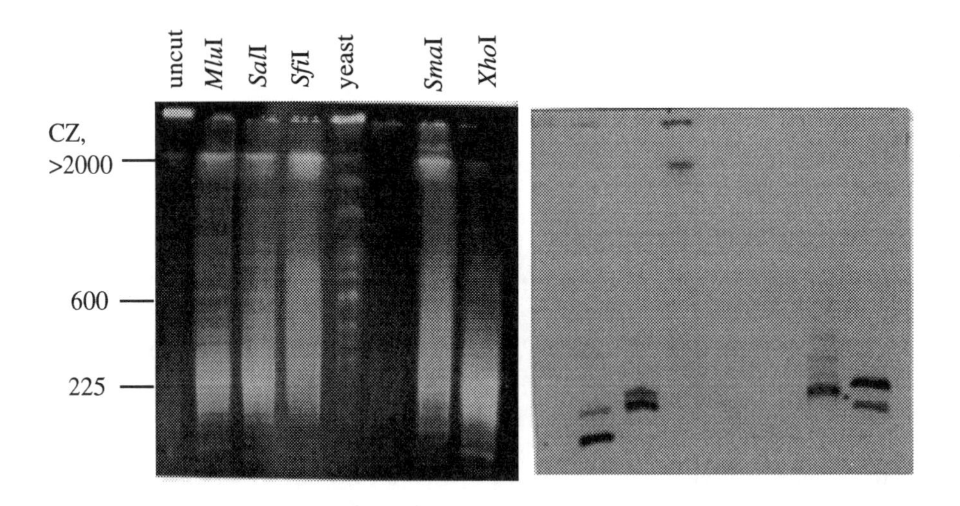

Figure 2. Digestion of agarose embedded soybean DNA with several restriction enzymes, and Southern hybridization with a marker from the soybean RFLP map. Conditions of electrophoresis were a 50s pulse time for 15 h followed by a 90s pulse time for 9 h in 0.5 x TBE, 200V, at 14 °C. A CHEF-DRII aparatus was used in all experiments. The ethidium bromide stained gel shows that the majority of DNA that is not cut with a restriction enzyme remains in the well or in the compression zone. The appearance of characteristic discrete bands in the other lanes is evidence of complete digestion. Hybridization to the well and the compression zone of the lane containing the *Sfi*I digest shows that the *Sfi*I restriction fragments detected by the marker are too large to be resolved under these conditions of electrophoresis.

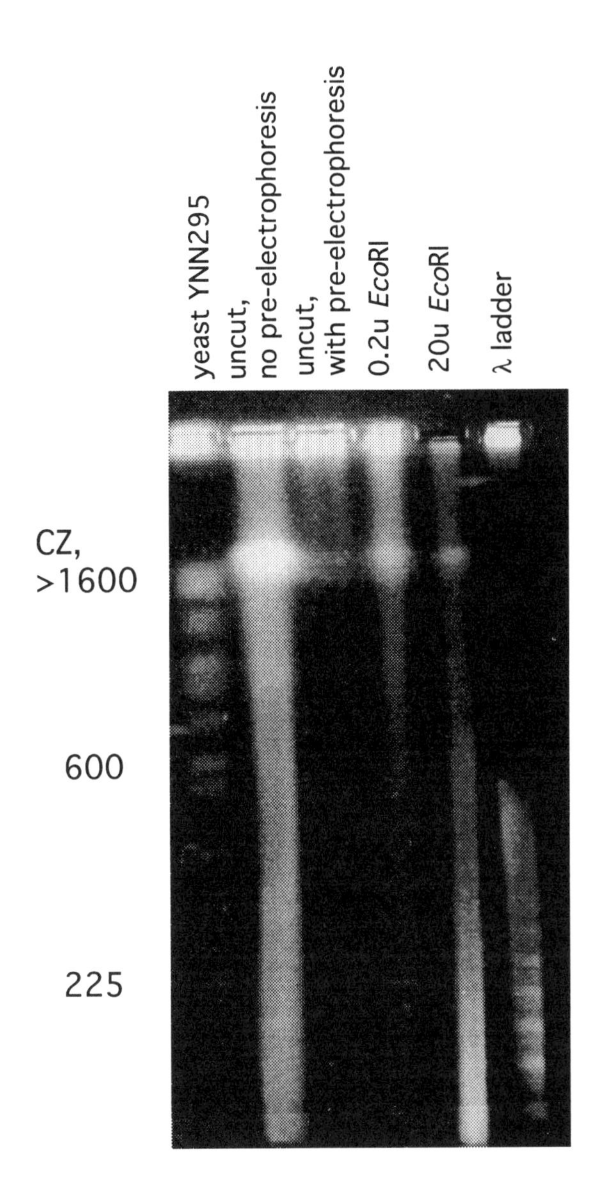

Figure 3. Efficiency of pre-electrophoresis for removal of degraded DNA from plugs, and partial digestion. Without pre-electrophoresis, this batch of plugs contained a considerable amount of degraded DNA which trails along the whole lane. After pre-electrophoresis, this fraction is almost completely removed as demonstrated by the retention of most of the remaining DNA in the well. The sensitivity of the DNA to partial digestion is shown by the fact that as little as 0.2U of *Eco*RI is sufficient to produce visible digestion. Conditions for pre-electrophoresis are described in the text. Conditions for the gel shown were 20-70s switch interval ramped over 20 hours in 0.25 x TBE at 14 °C.

Pre-electrophoresis

Highest quality megabase DNA is necessary for both cloning and long range mapping (fragments larger than 1 Mb). However, plugs almost always contain a fraction of relatively low molecular weight DNA (50-300 kb). The presence of this sheared DNA in plugs will yield a large number of fragments with at least one end that is not suitable for cloning. In mapping, it generates undesirable smearing of the signal. The degraded DNA may be removed by pre-electrophoresis prior to complete or partial digestion and cloning, leaving only very high molecular weight DNA in the plugs (30). The efficiency of the procedure can be improved by equilibrating the plugs with low ionic strength buffer before loading (31).

1. DNA plugs are treated with TE buffer and PMSF as described above.
2. The plugs are placed in a long slot of a 1% gel.
3. Running conditions are 60 s pulse time, 4 hours, in 0.5x TBE, 14 °C.
4. The DNA plugs are retrieved from the well and rinsed two times with TE prior to partial digestion and ligation. Modifications of conditions for partial digestion exist. One may use limiting amounts of enzyme for a set length of time, or perform a time course for incubation with a set amount of enzyme. The amount of Mg^{2+} may be titrated after incubating the plugs in restriction buffer lacking Mg^{2+} for 4 h on ice (32). Finally, different ratios of *Eco*RI and *Eco*RI methylase may be employed in a compromise buffer to achieve the desired degree of digestion (33). A typical result with soybean DNA is shown in Figure 3. In this case, the different number of units of *Eco*RI were allowed to diffuse into the plugs (30-40 µl each, in 50 µl *Eco*RI buffer containing Mg^{2+}) for 30 min on ice. The digests were then placed at 37 °C and terminated after 1 h by the addition of 100 µl 50 mM EDTA containing 1 mg/ml proteinase K. The removal of degraded DNA from the original plugs permits a more accurate estimation of the appropriate conditions for partial digestion.

Second Dimension

1. For two dimensional electrophoresis, twice as much DNA (at the same concentration) is used in the first dimension PFGE to improve the signal intensity on hybridization.
2. After staining with ethidium bromide and photographing, lanes of fractionated DNA are cut out of the gel and equilibrated on ice in slender glass tubes with the buffer appropriate for the enzyme to be used in the second dimension for one hour.
3. About 200 units of the enzyme are added and allowed to diffuse into the slice for an additional hour before placing the digests at 37 °C overnight.
4. The slices are briefly (15-20 min) equilibrated with running buffer.
5. They are then placed in a casting tray perpendicular to the direction of electrophoresis and embedded in a large gel. A normal plug of DNA not fractionated by PFGE and digested with the same enzymes is loaded alongside the slice in order to be able to identify individual copies of markers, members of a gene family, or repeated elements.
6. Following electrophoresis, the gel is transferred and blotted using conventional procedures.

A typical result showing how second dimension electrophoresis was used to assign individual copies of a marker in the soybean genome to separate bands produced by *Sfi*I digestion on a PFGE Southern is shown in Figure 4.

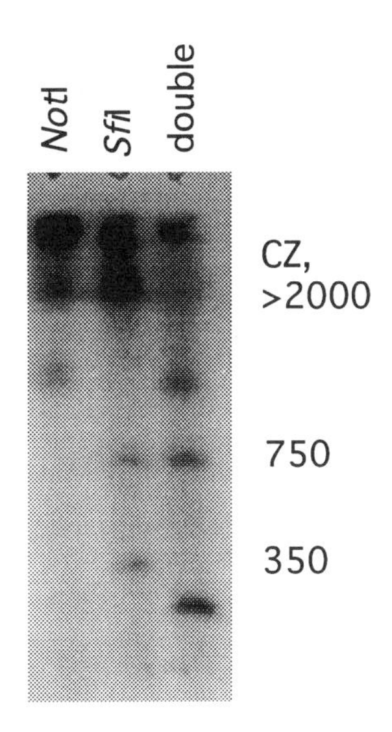

a. Southern blot of PFGE resolved digests of soybean DNA

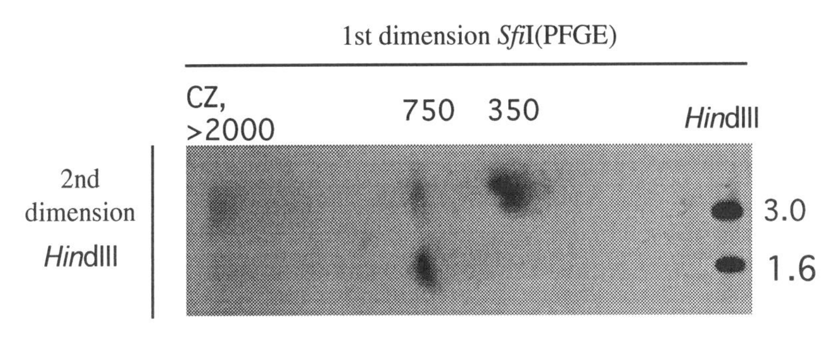

b. Southern blot of second dimension electrophoresis

Figure 4. Two dimensional electrophoresis of soybean DNA. A soybean molecular marker which has two copies in the genome, one polymorphic and one monomorphic, detected two distinct bands in *Sfi*I digests, as well as hybridizing to the compression zone (a). To assign the monomorphic and polymorphic copies of the marker (corresponding to 3.0 and 1.6 kb *Hin*dIII fragments, respectively) to the *Sfi*I bands, a second dimension electrophoresis was performed as described in the text (b). The result shows that only the monomorphic copy of the marker resides on a 350 kb *Sfi*I fragment, as well as on a partial product at 750 kb. The polymorphic copy of the marker is situated on a separate 750 kb *Sfi*I fragment.

Preparative Gels

Preparative PFGE should be run in low-melting point agarose which gives slightly better resolution than regular agarose. If DNA is intended for cloning, precautions should be taken not to stain the bands with ethidium. Instead, side slices of the gel are stained and the location of the desired band is marked by making notches in the gel. If the preparation will be labeled, the whole gel can be stained. Figure 5 shows an example of preparative isolation of several YACs containing soybean DNA for fluorescent *in situ* hybridization (FISH).

Figure 5. Preparative electrophoresis of soybean YACs. YACs were produced and plugs containing chromosomal DNA prepared as described (31, 2). 150 ml plugs were loaded on a 1% low melting point agarose gel, and a 40s pulse time was applied for 20 h in 0.5 x TBE at 14 °C. The entire gel was stained in this case because it was not necessary for the DNA to be intact for the application for which it was intended.

References

1. Schwartz, D.C., Saffran, W., Welsh, J., Haas, R., Goldberg, M. and Cantor, C. 1982. New techniques for purifying large DNAs and studying their properties and packaging. Cold Spring Harbor Symp. Quant. Biol. 47: 189-195.
2. Birren, B. and Lai, E. 1993. Pulsed field gel electrophoresis: a practical guide. Academic Press, San Diego.
3. Pulsed-field gel electrophoresis. 1992. M. Burmeister and L. Ulanovsky, eds. Humana Press, Totowa, NJ.
4. Monaco, A.P. and Larin, Z. 1994. YACs, BACs, PACs and MACs: artificial chromosomes as research tools. Trends in Biotechnol. 12: 280-286.
5. Moore, G., Gale, M.D., Kurata, N. and Flavell, R.B. 1993. Molecular analysis of small grain cereal genomes: current status and prospectus. Bio/Technology. 11: 584-589.
6. Civardi, L., Xia, Y., Edwards, K.J., Schnable, P.S. and Nikolau, B.J. 1994. The relationship between genetic and physical distances in the cloned *a1-sh2* interval of the *Zea mays* L. genome. Proc Natl. Acad. Sci. USA. 91: 8268-8272.
7. Warburton, P.E. and Willard, H.F. 1990. Genomic analysis of sequence variation in tandemly repeated DNA in the human genome. J. Mol. Biol. 216: 3-16.
8. Kolchinsky, A. and Gresshoff, P.M. 1994. Plant telomeres as molecular markers. In: Plant genome analysis. P.M. Gresshoff, ed. CRC Press, Roca Baton, Florida.
9. Ambrose, H.J., Byrd, P.J., McConville, C.M., Cooper, P.R., Ctankovic, T., Riley, J.H., Shiloh, Y., McNamara, J.O., Fukao, T. and Taylor, A.M.R. 1994. A physical map across chromosome 11q22-q23 containing the major locus for Ataxia Telangiectasia. Genomics. 21: 612-619.

10. Southard-Smith, M., Pierce, J.C., MacDonald, R.J. 1994. Physical mapping of the rat tissue kallikrein family in two gene clusters by analysis of P1 bacterophage clones. Genomics. 22: 404-417.
11. Ferrin, L.J. and Camerini-Otero, D. 1991. Selective cleavage of human DNA: RecA-assisted restriction endonuclease (RARE) cleavage. Science. 254: 1494-1497.
12. Ferrin, L.J. and Camerini-Otero, D. 1994. Long-range mapping of gaps and telomeres with Rec-assisted restriction endonuclease (RARE) cleavage. Nature Genet. 6: 379-383.
13. Negorev, D.G., Macina, R.A., Spais, C., Ruthig, L.A., Hu, X.-L. and Riethman, H.C. 1994. Physical analysis of the terminal 270 Kb of DNA from human chromosome 1q. Genomics. 22: 569-578.
14. Funke, R., Kolchinsky, A., and Gresshoff, P.M. 1993. Physical mapping of a region in the soybean genome containing duplicated sequences. Plant Mol. Biol. 22: 437-446.
15. Larsen, F., Gundersen, G., Lopez, R. and Prydz, H. 1992. CpG islands as gene markers in the human genome. Genomics. 13: 1095-1107.
16. Coleman, M., Németh, A.H., Campbell, L., Raut, C.P., Weissenbach, J. and Davies, K.E. 1994. A 1.8-Mb YAC contig in Xp11.23: identification of CpG islands and physical mapping of CA repeats in a region of high gene density. Genomics. 21: 337-343.
17. Verdoes, J.C., Calil, M.R., Punt, P.J., Debets, F., Swart, K., Stouthamer, A.H. and van den Hondel, C.A.M.J.J. 1994. The complete karyotype of *Aspergillus niger*: the use of introduced electrophoretic mobility variation of chromosomes for gene assignment studies. Mol. Gen. Genet. 244: 75-80.
18. Kolchinsky, A. Funke, R. P. and Gresshoff, P.M. 1993. DAF-amplified fragments can be used as markers for DNA from pulse field gels. Biotechniques. 14: 400-403.
19. Denton, P.H., Cullen, J.B., Loeb, D., Lucas, A. and Nunes, K. 1994. Partitioned PFGE-PCR: a new method for pulsed-field mapping for SRS and microsatellites. Nucl. Acids Res. 22: 1776-1777.
20. Gaensler, K.M.L., Kitamura, M. and Kan, Y.W. 1993. Germ-line transmission and developmental regulation of a 150 kb yeast artificial chromosome containing the human beta-globin locus in transgenic mice. Proc. Natl. Acad. Sci. USA. 90: 11381-11385.
21. Peterson, K.R., Zitnik, G., and Huxley, C. 1993. Use of YACs for studying control of gene expression: correct regulation of the genes of a human beta-globin locus YAC after transfer to mouse erythroleukemia line. Proc. Natl. Acad. Sci. USA. 90: 11207-11211.
22. Lamb, B.T., Sisodia, S.S. and Lawler, A.M. 1993. Introduction and expression of the 400 kilobase precursor amyloid protein gene in transgenic mice. Nature Genet. 5: 22-26.
23. Demmer, L.A. and Chaplin, D.D. 1993. Simultaneous transfer of four functional genes from the HLA class II region into mammalian cells by fusion with yeast spheroplasts carrying an artificial chromosome. J. Immunol. 150: 5371.
24. Ketner, G., Spencer, F., Tugendreich, S., Connelly, C. and Hieter, P. 1994. Efficient manipulation of the human adenovirus genome as an infectious yeast artificial chromosome clone 1994. Proc. Natl. Acad. Sci. USA. 91: 6186-6190.

25. Schwartz, D.C. and Cantor, C.R. 1984. Separation of yeast chromosome-sized fragments by pulsed field gradient gel electrophoresis. Cell. 37: 67-75.
26. Wing, R.A., Rastogi, V.K., Zhang, H.B., Paterson, A.H. and Tanksley, S.D. 1993. An improved method of plant DNA megabase isolation in agarose microbeads suitable for physical mapping and YAC cloning. Plant J. 4: 893-898.
27. Liu, Y.-G. and Whittier, R.F. 1994. Rapid preparation of megabase plant DNA from nuclei in agarose plugs and microbeads. Nucl. Acids Res. 22: 2168-2169.
28. Doggett, N.A., Smith, C.L. and Cantor, C.R. 1992. The effect of DNA concentration on mobility in PFGE. Nucl. Acids Res. 4: 859-864.
29. Walter, M.A., Surti, U., Hofker, M.H. and Cox, D.W. 1990. The physical organization of the human immunoglobulin heavy chain gene locus. EMBO J. 9: 3303-3313.
30. Edwards, K.J., Thompson, H., Edwards, D., de Saizieu, A., Sparks, C., Thompson, J.A., Greenland, A.J., Eyers, M. and Scuch, W. 1992. Construction and characterization of a YAC library containing three haploid maize genome equivalent. Plant Mol. Biol. 22: 437-446.
31. Funke, R., Kolchinsky, A. and Gresshoff, P.M. 1994. High EDTA concentration causes entrapment of small DNA molecules in the CZ of PFGE, resulting in smaller than expected insert sizes of YACs. Nucl. Acids Res. 22: 2708-2709.
32. Albertsen, H.M., Abderrahim, H., Cann, H.M., Dausset, J., LePaslier, D. and Cohen, D. 1990. Construction and characterization of a YAC library containing seven haploid human genome equivalents. Proc. Natl. Acad. Sci. USA. 87: 4256-4260.
33. Larin, Z., Monaco, A.P. and Lehrach, H. 1991. YAC libraries containing large inserts from mouse and human DNA. Proc. Natl. Acad. Sci. USA. 88: 4123-4127.

From: *Molecular Biology: Current Innovations and Future Trends.*
ISBN 1-898486-01-8 ©1995 Horizon Scientific Press, Wymondham, U.K.

6

AUTOMATED DNA HYBRIDIZATION AND DETECTION

Stephan Beck

Abstract

The specific identification of DNA molecules by hybridization with probes complementary to their target sequences has wide ranging applications in research, healthcare and industry. The simplicity of the hybridization procedure in combination with advanced detection systems is well suited for processing large numbers of samples and, with various genome projects gathering pace, DNA hybridization and detection have now become a major focus for automation. Here, the current status and future trends in this field of research will be reviewed and potential applications will be discussed using genome analysis as example.

Introduction

The technique of DNA hybridization was first reported in 1961 (1) but its breakthrough, at least for membrane hybridization, came only in 1975 when it was combined with a method to transfer or blot electrophoretically separated DNA fragments from a gel onto a solid support - now known as "Southern Blotting" (2). The principle of DNA hybridization is based on the ability of two complementary DNA strands to form a double helix or "hybrid" which is held together by hydrogen bonding. The same principle is equally valid for DNA-RNA hybridization. By labeling one strand which subsequently is used as "probe" in the hybridization, the complementary strand or "target" can be identified with greatest specificity.

The efficiency and kinetics of DNA hybridization depend upon various parameters such as the buffer composition (pH, ionic strength), the hybridization/wash temperature and the probe itself (single/double stranded, length, base composition). These and further hybridization parameters have already been extensively reviewed (3, 4) and will not be covered again here. Before we turn our attention to the subject of this review, automation of hybridization and detection, I would just like to mention a few words about the various labels which are currently used for detection. They can be divided into two categories: (i) radioactive labels (e.g. ^{32}P, ^{33}P, ^{35}S, ^{3}H) and increasingly (ii) non-radioactive labels. The most popular ones of the latter category are based on enzyme-catalysed colourimetric staining (5), chemiluminescence (6) or fluorescence

(7). Their chemical mechanisms and labeling procedures have also been subject to several previous reviews (8-10).

For any process to be automated it is most helpful if it can be broken into clearly defined, individual steps. This can easily be achieved for DNA hybridization as shown in Figure 1. Virtually all hybridization strategies involve the same four steps: Prehybridization (to block nonspecific probe binding sites), hybridization, wash and detection. Additional modification steps are needed for some strategies (e.g. using colourimetric or chemiluminescent detection) and probe stripping steps are required if the same support needs to be hybridized multiple times. The wide spread use and apparent simplicity of the entire process clearly demonstrate the need for automation of the hybridization technology. In this chapter I will review the state-of-the-art instrumentation in particular for membrane, *in situ* and magnetic bead hybridization and I will also discuss some of the emerging future trends.

Past and Current Developments

Membrane Hybridization

In membrane hybridization, the target DNA or RNA is extracted and purified prior to immobilisation onto a solid support. In some procedures (e.g. plaque lifts) the purification is carried out after the immobilisation, directly on the solid support. In the past, nitrocellulose was the most widely used solid support but today, a wide range of nylon membranes are available which have superior chemical and mechanical properties. To a large extent, the process of immobilisation has already been automated by the use of robotic workstations for DNA spotting (11) and, for example, direct blotting electrophoresis (12) for DNA transfer from gels onto membranes. Afterwards, the membranes are usually baked and irradiated with ultraviolet light (uv-crosslink) in order to bind the DNA/RNA covalently to the membrane.

The very first automated hybridization system was described in 1986 (13). Certainly at that time, it was an ambitious attempt to prove that gel electrophoresis, hybridization and detection - essentially Southern blotting - could not only be automated but also integrated into a single system. The resulting system was of enormous proportions consisting of a robot arm moving samples between electrophoresis, hybridization and detection modules (14, 15). The system may well have been a bit ahead of its time but today, it is technologically outdated particularly the detector module which is based on an array of Geiger counters.

Subsequent developers abandoned the fully integrated approach in favour of stand-alone systems optimized for one specific application. Among these systems is an automated probing device designed to assist multiplex DNA sequencing (16). In the multiplex strategy DNA sequence images are obtained by consecutive hybridizations of the same membrane with specific tag-oligonucleotides. In this particular device each membrane is heatsealed into a special pouch which is then fixed horizontally onto a shelf and connected to the fluidic system. Prehybridization, hybridization, washes and detection are all carried out in a thermo-controlled and light-tight box with can hold multiple shelfs simultaneously. Solvent delivery and removal is achieved by pumps and a series of valves which allow individual shelves to be operated independently.

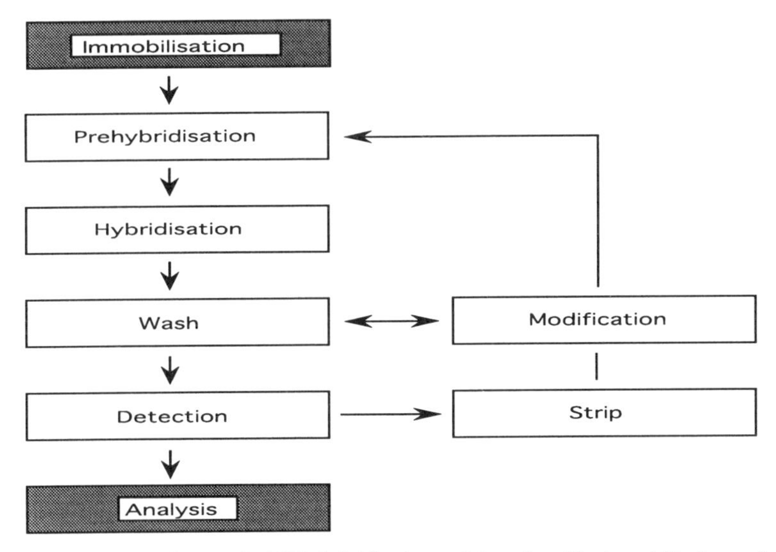

Figure 1. General flowdiagram for DNA hybridization and detection. The immobilisation and analysis steps are shaded because they are not strictly part of of the actual hybridization/detection procedure.

The waste container has been specially modified to concentrate all liquid waste 100-fold in order to reduce disposal cost, in particular of radioactive waste. For the detection, X-ray films are manually placed on each shelf and good contact between the films and the membranes is ensured by inflating air bags directly over the films. The computer control of the individual steps allows sufficient experimental freedom for the device to be suitable as general purpose hybridizer. More recent developments have replaced the manual, radioactive detection by automated, enzyme-linked fluorescence detection (7). In this system the hybridized membrane is scanned with an argon laser and the emitted fluorescence is detected with a cryogenically cooled charge-coupled device (CCD) camera.

Library screening is another application which relies on hybridization. Usually, such libraries are spotted as high density arrays or grids onto nylon membranes for repeated hybridization cycles (17). In our laboratory, we have developed a system which is capable of automated hybridization and detection of such gridded libraries (18). A schematic outline of our system is shown in Figure 2. It consists of a home-made hybridizer and a commercially available detection unit both operated and controlled by a PC computer. The hybridizer consists of two major parts, the membrane processing cassette and the fluidic system. The processing cassette is thermo-controlled and can hold multiple membranes simultaneously. The temperature within the cassette can be set freely and can be maintained to within +/- 1°C. The fluidic system consists of seven independent input channels, a collecting waste container and flow controlling valves. There is no pump required since the entire fluidic system is gravity driven. The hybridizer is fully compatible with non-radioactive detection chemistries such as enzyme-triggered chemiluminescence. The detection unit (BIQ, Cambridge Imaging, Cambridge, UK, see also ref. 19) consists of an image intensified CCD camera (which does not require cooling) and image acquisition/analysis software (BIOVIEW, Cambridge Imaging, Cambridge, UK). Similar to above, our system could be turned into a general purpose hybridizer with only minor modifications.

Unlike automated DNA synthesizers and DNA sequencers, automated hybridization/detection systems are not yet commercially available. However, the first automated membrane processor has just been introduced (model PR1000, Hoefer Scientific Instruments, San Francisco CA, USA). Operating only at ambient temperature and using roller bottles for low solvent consumption it is similar to a previously described processor which aims to automate the numerous incubation and wash steps required for most non-radioactive hybridization procedures (20).

The development of instrumentation suitable for the detection of hybridized membranes and other supports seems to have attracted more commercial interest. The Geiger counter array (15) described above has today been succeeded by multi-wire proportional counters (MWPC) or phosphorimagers for radioactive detection and CCDs for non-radioactive detection. MWPC counters combine quantitation of scintillation counters and area detection of X-ray films into a single instrument (21). They are capable of submillimeter spatial resolution for most of the common radiolabels. Commercially available MWPC counters which have been especially adapted to detect and analyse hybridizations are, for example, the Betascope 603 ($^{32-}$P only, Betagen Corporation, Waltham MA, USA, ref. 22, 23) and the InstantImager ($^{14-}$C, $^{35-}$S, $^{32-}$P, $^{33-}$P, $^{125-}$I, Canberra Packard, Meriden CT, USA). With over 200,000 solid state detectors the InstantImager can image and quantify an area of 20x24 cm which can be viewed in real time (24). Phosphorimagers are based on the ability of photostimulable phosphor crystals to store a fraction of the absorbed incident energy from irradiation with, for example, $^{32-}$P, $^{33-}$P, $^{35-}$S or $^{14-}$C (25, 26). Erasable imaging plates coated with such phosphor crystals emit luminescence proportional to the absorbed radiation energy (for example from a hybridized membrane) when stimulated by visible or infrared radiation. Fully integrated phosphor imaging systems are available from Fuji Photo Film Co. Ltd, Tokyo, Japan (Model FUJIX BAS) and Molecular Dynamics, Sunnyvale CA, USA (Model 400A, see also ref. 26). For non-radioactive detection especially of bio- and chemiluminescence, CCD cameras have become the detector of choice. A solid-state CCD detector usually is a silicone chip (1 cm^2) subdivided by electrodes into photosensitive elements called pixels. A suitable CCD contains 512x512 pixels and has a dynamic range between 450-700 nm of about 10^4 (max signal/min signal). When photons are absorbed by such pixels, electron-hole pairs are created which are converted into a signal by a process termed charge-coupling (27). The image intensified BIQ system (Cambridge Imaging, Cambridge, UK), for example, allows quantitative detection of micro and macro samples in real time (19). Photomultipliers can, of course, be used as alternative to CCD cameras especially for fluorescent detection. Such a system using laser-excited fluorescence has been developed for the detection of entire gels and hybridized membranes (28).

In Situ Hybridization

In contrast to membrane hybridization, the histological cell structure is maintained in *in situ* hybridization. It allows the precise localization and identification of target DNA and RNA within tissues and individual cells and has widespread applications in immuno-histochemistry, pathology and cytogenetics (29). A specialized *in situ* technique, known as fluorescence *in situ* hybridization or FISH has proven to be particularly useful for genome and gene mapping by allowing the analysis of chromosomal preparations (30).

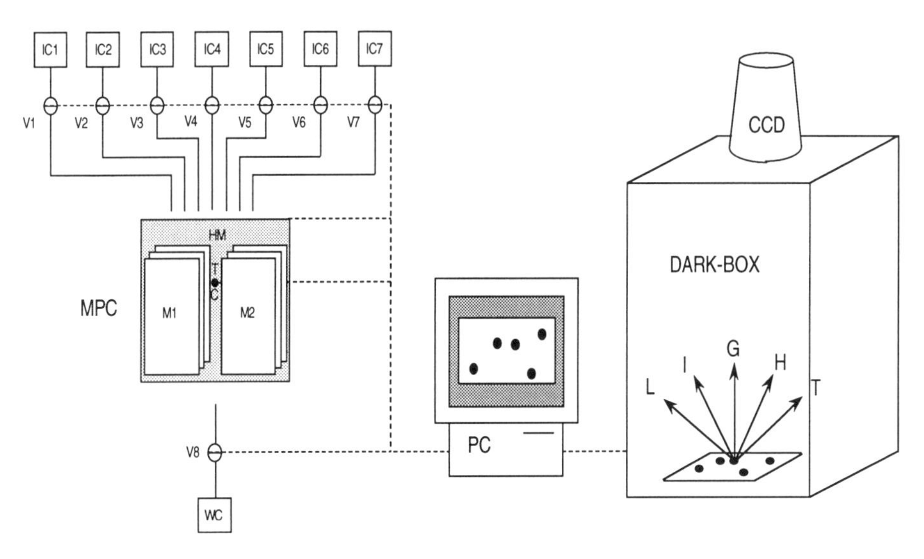

Figure 2. Example of automated DNA hybridization with CCD detection. IC indicates input container; V, electric solenoid valve; MPC, membrane processing cassette; HM, heating matt; M, membrane; TC, thermocouple; PC, computer; WC, waste container and CCD, charge coupled device. For more details see text and reference 18.

Despite the differences to membrane hybridization, *in situ* hybridization follows the same steps outlined in Figure 1. Instead of membranes, glass slides are used as solid supports to immobilize tissue sections, cells or chromosome spreads.

Several automated systems have been developed for *in-situ* hybridization. The first was described in 1988 using capillary action for reagent delivery and a blotting pad for reagent removal in conjunction with a robotic workstation (31). Multiple glass slides arranged to form capillary gaps are moved by a robot arm between reagent containers, thermo-controlled incubation chambers and a blotting pad. The system is suitable for colorimetric detection and is commercially available in a manual version (MicroProbe System, Fisher Scientific Inc., Pittsburg PA, USA, see also ref 32) and an automated version (Code-On System, Instrumentation Laboratories, Lexington MA, USA). A second system follows a similar approach but, instead of a robot arm, it uses a valve-controlled fluidic system driven by a pump for reagent delivery and vacuum for reagent removal (33). Recently, two compact benchtop systems were introduced. The semi-automatic OmniSlide system (Hybaid Ltd., Teddington, UK) integrates thermal cycling into the hybridization step and in conjunction with off-line wash modules it allows for more sensitive detection (34). In the Ventana system (Ventana Medical System Inc., Tucson, AZ, USA) the sample slides (up to 40) and reagents (up to 25) are arranged on carousels and every process is monitored by an integrated bar code system for error free operation. In summary, the current degree and availability of automation appears to be more advanced for *in situ* hybridization than for membrane hybridization.

Magnetic Bead Hybridization

The coupling of hybridization with magnetic bead separation has created a versatile new technique for the identification and isolation of biomolecules (35). For DNA/

RNA capture a biotinylated probe is normally used for the hybridization and after the addition of streptavidin-coated magnetic beads the hybrid can be selectively isolated from a complex mixture with the help of a magnet. Such beads are available from Dynal (Dynabeads, Dynal, Oslo, Norway) and other suppliers. The entire process can be automated easily by using a laboratory robot such as the 'PolySeq Biomek1000' (Beckman Instruments Inc., Fullerton CA, USA; 43), the 'Vistra DNA Labstation' (Amersham International plc, Amersham, UK) or the 'CATALYST Mag Bead Station' (Applied Biosystems Inc, Foster City CA, USA). All three robots have an integrated magnetic separator.

Future Trends

Apart from optimizing existing technology such as the systems discussed above, the key to future developments in hybridization and detection technology most likely lies in miniaturization. While current technology operates in the ml volume range and, at best, in the attomol sensitivity range, future technologies are very likely to operate in the nl-µl range with a sensitivity in the 1-100 molecule range. Of the emerging trends to achieve this goal two approaches currently seem to be most promising: the development of hybridization chips and biosensors.

Borrowed from the world of computer technology, several successful attempts have been made to immobilize or synthesize high-density oligonucleotide arrays on a solid support (e.g. glass) which subsequently can be used as a hybridization chip (36-38). Using photolithographic masking for instance, chips with 25 µm array resolution have been achieved and successfully hybridized (38). At this density all possible

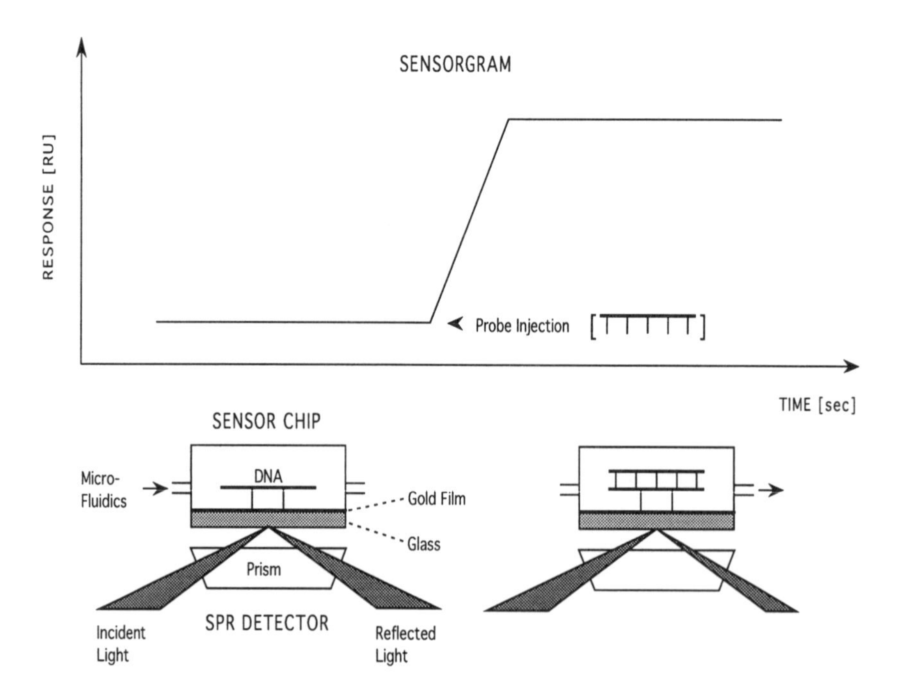

Figure 3. Schematic outline of DNA hybridization using surface plasmon resonance (SPR).

octanucleotides ($4^8 = 65536$) would fit in an array smaller than 1 cm^2. In order to increase detection speed and sensitivity oligonucleotide arrays have even been created directly on the surface of a charge coupled device (CCD) which results in much reduced detection times (39). In summary, hybridization chip technology in conjunction with microelectronics holds high hopes for a wide range of applications such as DNA diagnostics and DNA sequencing (40).

The second approach is based on an optical phenomenon which is known as surface plasmon resonance or SPR (41). At a certain angle, incident light is totally reflected at the interface of two media with different refractive indices such as glass and buffer. If a thin metal film (e.g. gold) is inserted at the interface, its free electron clouds can couple and resonate with the incident light to form plasmons. The resonance absorbs some energy from the incident light which results in a measurable drop in reflected incident light intensity which correlates to changes in the refractive indices. Therefore, SPR is ideally suited to analyse biomolecular interactions including DNA hybridization. In fact, using SPR and the BIAcore system (Pharmacia Biosensor AB, Uppsala, Sweden) it is possible to analyse the kinetics of DNA hybridization in real time without any label and even detect mismatches (42). The BIAcore is the first instrument of its kind consisting of a SPR detector, a microfluidic system, biosensor chips with different, functionalized surface groups for efficient immobilization and integrated application software. A simplified schematic for DNA hybridization using SPR is shown in Figure 3.

Acknowledgment

I would like to thank D. Englert and R. Zeheb for providing information prior to publication and H. Lehrach and N. Spurr for critical reading of the manuscript.

References

1. Hall, B. D. and Spiegelman, S. 1961. Sequence complementarity of T2-DNA and T2-specific RNA. Proc. Natl. Acad. Sci. USA. 47: 137- 146.
2. Southern, E. 1975. Detection of specific sequences among DNA fragments separated by gel electrophoresis. J. Mol. Biol. 98: 503-517.
3. Meinkoth, J. and Wahl, G. 1984. Hybridization of nucleic acids immobilized on solid supports. Anal. Biochem. 138: 267-284.
4. Hames, B.D. and Higgins, S.J. 1985. Nucleic acid hybridization: a practical approach. IRL Press, Oxford.
5. Leary, J. J., Brigati, D. J. and Ward, D. C. 1983. Rapid and sensitive colorimetric method for visualizing biotin-labeled DNA probes hybridized to DNA or RNA immobilized on nitrocellulose: Bio-blots. Proc. Natl. Acad. Sci. USA 80: 4045-4049.
6. Beck, S., O'Keeffe, T., Coull, J. M. and Köster, H. 1989. Chemiluminescent detection of DNA: applications for DNA sequencing and hybridization. Nucleic Acids Res. 17: 5115- 5123.

7. Cherry, J. L., Young, H., Di Sera, L. J., Ferguson, F. M., Kimball, A. W., Dunn, D. M., Gesteland, R. F. and Weiss, R. B. 1994. Enzyme-linked fluorescent detection for automated multiplex DNA sequencing. Genomics 20: 68-74.
8. Wilchek, M. and Bayer, E. A. 1988. The avidin-biotin complex in bioanalytical applications. Anal. Biochem. 171: 1-32.
9. Kessler, K., Höltke, H.-J., Seibl, R., Burg, J. and Mühlegger, K. 1990. Non-radioactive labeling and detection of nucleic acids. Biol. Chem. Hoppe-Seyler 371: 917-965.
10. Beck, S. and Köster, H. 1990. Applications of dioxetane chemiluminescent probes to molecular biology. Anal. Chem. 62: 2258-2270.
11. Maier, E., Meier-Ewert, S., Ahmadi, A., Curtis, J. and Lehrach, H. 1994. Application of robotic technology to automated sequence fingerprint analysis by oligonucleotide hybridization. J. Biotechnol. 35: 191-203.
12. Beck, S. 1993. Direct blotting electrophoresis. In: DNA Sequencing Protocols. H.G. Griffin, and A.M. Griffin, eds. Humana Press, Totowa, NJ. p. 219-223.
13. Gersten, D. M., Zapolski, E. J., Golab, T., Buas, M. and Ledley, R. S. 1986. Computer controlled DNA electrophoresis and hybridization. In: Electrophoresis '86. M.J. Dunn, ed. VCH, Weinheim. p. 187-190.
14. Zapolski, E. J., Buas, M., Golab, T., Ledley, R. S. and Gersten, D. M. 1987. A system for automated DNA electrophoresis, molecular hybridization and electronic detection: I. Electrophoresis and hybridization. Electrophoresis 8: 255-261.
15. Zapolski, E. J. and Gersten, D. M. 1989. A system for automated DNA electrophoresis, molecular hybridization and electronic detection: II. Electronic detection. Electrophoresis 10: 1-6.
16. Church, G. M. and Kieffer-Higgins, S. 1989. Automated probing device for multiplex sequencing. European Patent No. 0 303 459 A2.
17. Lehrach, H., Drmanac, R., Hoheisel, J., Larin, Z., Lennon, G., Monaco, A. P., Nizetic, D., Zehetner, G. and Poustka, A. 1990. Hybridization fingerprinting in genome mapping and sequencing. In: Genome Analysis Vol. 1. Cold Spring Harbor Laboratory Press, Cold Spring Harbor, New York. p. 39-81.
18. Alderton, R. P., Kitau, J. and Beck, S. 1994. Automated DNA hybridization. Anal. Biochem. 218: 98-102.
19. Hooper, C. E. and Ansorge, R. E. 1990. Quantitative luminescence imaging in the biosciences using the CCD camera: analysis of macro and micro samples. Trends in Anal. Chem. 9: 269-277.
20. Richterich, P., Heller, C., Wurst, H. and Pohl, F.M. 1989. DNA sequencing with direct blotting electrophoresis and colorimetric detection. Biotechniques 7: 52-59.
21. Bateman, J. E. 1990. Quantitative autoradiographic imaging using gas-counter technology. Electrophoresis 11: 367-375.
22. Sullivan, D.E., Auron, P.E., Quigley, G.J., Watkins, P., Stanchfield, J.E. and Bolon, C. 1987. The nucleic acid blot analyser: I: High speed imaging and quantitation of ^{32}P- labeled probes. Biotechniques 5: 672-678.
23. Auron, P. E., Sullivan, D., Fenton, M. J., Clark, B. D., Cole, E. S., Galson, D. L., Peters, L. and Teller, D. 1988. The nucleic acid blot analyzer II. Analyze, an image analysis software package for molecular biology. Biotechniques 6: 347- 353.
24. Englert, D., Roessler, N., Jeavons, A. and Fairless, S. 1994. Microchannel array detector for quantitative electronic radioautography. Cellular and Molecular Biology: in press.

25. Amemiya, Y. and Miyahara J. 1988. Imaging plate illuminates many fields. Nature 336: 89-90.
26. Johnston, R. F., Picket, S. C. and Barker, D. L. 1990. Autoradiography using storage phosphor technology. Electrophoresis 11: 355-360.
27. Mackay, C. D. 1986. Charge-coupled devices in astronomy. Ann. Rev. Astrom. Astrophys. 24: 255-283.
28. Ishino, Y., Mineno, J., Inoue, T., Fujimiya, H., Yamamoto, K., Tamura, T., Homma, M., Tanaka, K. and Kato, I. 1992. Practical aplications in molecular biology of sensitive fluorescence detection by a laser-excited fluorescence image analyzer. Biotechniques 13: 936-943.
29. Herrington, C. S. and McGee, J. O'D. 1992. Diagnostic molecular pathology: a practical approach. IRL Press, Oxford.
30. Korenberg, J. R., Yang-Feng, T., Schreck, R. and Chen, X. N. 1992. Using fluorescence in situ hybridization (FISH) in genome mapping. Trends in Biotechnol. 10: 27-32.
31. Unger, E. R., Brigati, D. J., Chenggis, M. L., Budgeon, L. R., Koebler, D., Cuomo, C. and Kennedy, T. 1988. Automation of *in situ* hybridization: application of the capillary action robotic workstation. J. Histotechnol. 11: 253-258.
32. Chenggis, M. L. and Unger, E. R. 1993. Application to a manual capillary action workstation to colorimetric *in situ* hybridization. J. Histotechnol. 16: 33-38.
33. Takahashi, T. and Ishiguro, K. 1991. Development of an automatic maschine for *in situ* hybridization and immunohistochemistry. Anal. Biochem. 196: 390-402.
34. Goden, J. and Lawson, D. 1994. Rapid chromosome identification by oligonucleotide-primed i*n situ* DNA synthesis (PRINS). Hum. Mol. Gen. 3: 931-936.
35. Haukanes, B.-I. and Kvam, C. 1993. Application of magnetic beads in bioassays. Biotechnol. 11: 60-63.
36. Khrapko, K.R., Lysov, Y.P., Khorlin, A.A., Ivanov, I.B., Yershov, G.M., Vasilenko, S.K., Florentiev, V.L. and Mirzabekov, A.D. 1991. A method for DNA sequencing by hybridization with oligonucleotide matrix. DNA Sequence 1: 375-388.
37. Maskos, U. and Southern, E.M. 1992. Oligonucleotide hybridizations on glass supports: a novel linker for oligonucleotide synthesis and hybridization properties of oligonucleotides synthesized *in vitro*. Nucl. Acids. Res. 20: 1679-1684.
38. Caviani Pease, A., Solas, D., Sullivan, E.J., Cronin, M.T., Holmes, C.P. and Fodor, S.P.A. 1994. Light-generated oligonucleotide arrays for rapid DNA sequence analysis. Proc. Natl. Acad. Sci. USA 91: 5022-5026.
39. Lamture, J. B., Beattie, K.L., Burke, B.E., Eggers, M.D., Ehrlich, D.J., Fowler, R., Holis, M.A., Kosicki, B.B., Reich, R.K., Smith, D.J., Varma, R.S. and Hogan, M.E. 1994. Direct detection of nucleic acid hybridization on the surface of a charge coupled device. Nucl. Acids Res. 22: 2121-2125.
40. Mirzabekov, A.D. 1994. DNA sequencing by hybridization - a megasequencing method and a diagnostic tool? Trends in Biotechnol. 12: 27-32.
41. Welford, K. 1991. Surface plasmon-polaritons and their uses. Opt. Quant. Electr. 23: 1-27.
42. Davis, S. 1994. Kinetic characterization of DNA hybridization. J. Biomolecular Interaction Analysis 1: 27.
43. Rolfs, A. and Weber, I. 1994. Fully-automated, non-radioactive solid-phase sequencing of genomic DNA obtained from PCR. Biotechniques. 17: 782-787.

From: *Molecular Biology: Current Innovations and Future Trends*.
ISBN 1-898486-01-8 ©1995 Horizon Scientific Press, Wymondham, U.K.

7

AN IMPROVED SUBTRACTIVE HYBRIDIZATION METHOD USING PHAGEMID VECTORS

Christian E. Gruber and Wu-Bo Li

Introduction

Subtractive hybridization is a valuable tool which can be used to identify and enrich genes that are differentially expressed in different cell types, different developmental stages, pathological conditions, or in response to stimuli (i.e. fibroblast factor). Subtractive hybridization is defined as the hybridization between two DNA (or RNA) populations that are closely related, and the removal of the hybridized sequences common to both tissues (or cells). Subsequently, the unhybridized sequences can be preserved as a subtracted cDNA library.

The specificity of the subtracted cDNA library is mainly dependent on the thoroughness of the hybridization reaction. The nucleic acid concentration, hybridization temperature, reaction time, salt concentration, fragment size and the sequence complexity determine the rate of the hybridization reaction. Usually, an excess of one nucleic acid population (driver) is used to accelerate the hybridization to the other nucleic acid population (target). When the driver concentration is greater than 20:1 over the target population, the kinetics of the second-order reaction of the bimolecular hybridization between the two populations could display a pseudo-first order reaction (1). The kinetics of the hybridization can be defined by the equation, $S=e^{-kC_0t}$ in which S is the fraction of the unhybridized sequences in the system, k is the empirical hybridization constant that is dependent on the complexity, C_0 is the concentration of the nucleic acid which should be equal to the driver concentration, and t is the hybridization time. Therefore, C_0t, the product of the concentration of the nucleic acid in the system and the hybridization time, can be used to determine the extent of the hybridization. In other words, the hybridization can be driven by increasing the concentration of driver sequences or by prolonging the time of hybridization. Theoretically, if the C_0t value reaches 1,000-3,000 M•second, the unhybridized common sequences remaining in the target population should be less than 1%.

The specificity of the subtractive hybridization is also dependent on the level of hybrid removal. A variety of methods are used to separate the unhybridized single-stranded (ss) sequences from the double-stranded (ds) hybrids. Although affinity chromatography using hydroxylapatite (HAP) columns is routinely used to separate the ss nucleic acid from ds sequences, the ss sequence can become degraded. Biotin-streptavidin/phenol extraction, is a more efficient method to remove hybridized sequences (2). In this method, the driver is biotin-labeled by a photochemical or enzymatic reaction. The biotinylated driver is then used to drive the hybridization reaction, followed by incubation of the reaction mixture with streptavidin, and phenol extraction. After phenol extraction, the hybrids and the excess driver are removed to the interface, and the unhybridized ss target sequences stay in the aqueous phase.

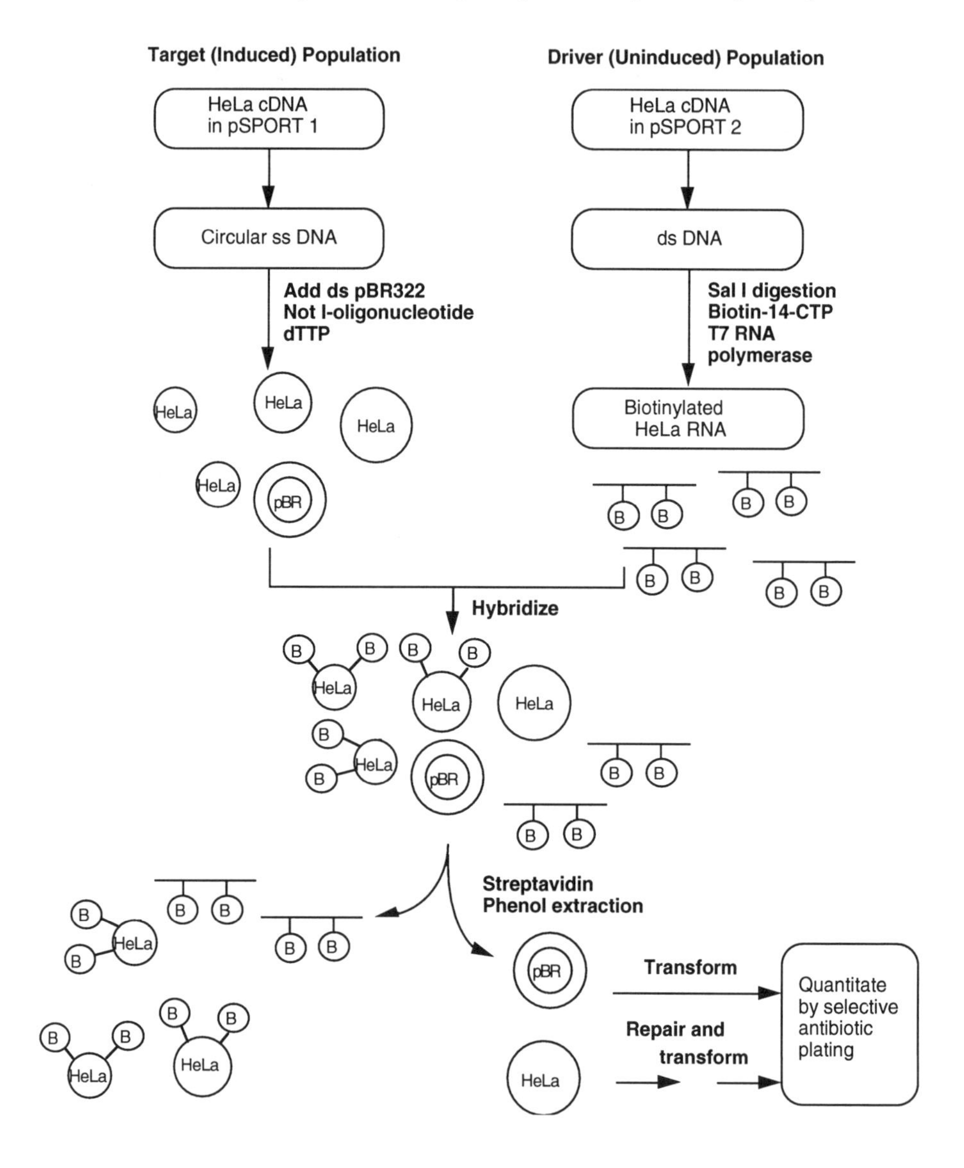

Figure 1. Schematic of the cDNA subtractive hybridization method. The ss DNA and ds DNA are represented by the single-circle and double-circle, respectively. The abbreviations represent: HeLa, cDNA from HeLa cells; pBR, plasmid pBR322; B, biotin attached to the driver. The ds pBR322 plasmid is added to the ss cDNA library. The HeLa ds driver plasmid is digested with restriction enzyme *Sal* I, and the biotinylated driver is prepared by *in vitro* transcription. The hybridization is performed at a C_0t value of 1000, followed by streptavidin binding and phenol/chloroform extraction. The subtracted ss cDNA is directly electroporated into *E. coli* competent cells or is repaired before the transformation. The transformants are then quantitated by plating onto the ampicillin and tetracycline plates.

Various driver and target nucleic acid populations can be used to produce subtracted cDNA libraries. For instance an RNA (target) can be hybridized to an RNA driver (or DNA driver) or a DNA (target) can be hybridized to an RNA driver (or DNA driver). In the first case, the target library is mRNA, and the driver is either complemetary RNA (cRNA) or DNA (3-5). The mRNA, remaining after the subtractive hybridization, is used as the template for cDNA synthesis and cloning. While successful, the mRNA target tends to become degraded during the subtraction making the subsequent library construction extremely difficult.

Target DNA produced from phagemid vector-cDNA libraries or PCR avoids this serious degradation problem. Duguid *et al.* (6) demonstrated successful subtractive hybridization using ss phagemid target and driver DNA populations. In this case, the ss phagemid driver was photobiotinylated, and hybridized with the ss phagemid target. After subtraction, the unhybridized ss DNA was directly used to transform competent cells. Another advantage is large amounts of driver mRNA, which can be difficult to obtain from some tissues or cells, are not required in this subtractive hybridization method. Rubenstein *et al.* (7) also described a method in which both target and driver libraries were directionally constructed in phagemid vectors. In their method, the unhybridized ss phagemid DNA target was converted to ds DNA after subtractive hybridization. This process effectively increases the tranformation efficiency of the small amount of unhybridized DNA. Although certainly improving upon past methods, these subtractive hybridization procedures required large amounts of highly purified ss DNA driver.

Recently, we described an improved and optimized method previously reported by Swaroop (8, 9). In this method, both driver and target cDNA libraries are directionally constructed in phagemid vectors. While other vectors can be used, our subtractive hybridization protocol (10, 11) uses the multifunctional cloning plasmids pSPORT 1 and pSPORT 2. The multiple cloning site (MCS) in pSPORT 2 is identical to the MCS of pSPORT 1, except the *Not* I-*Sal* I region of the pSPORT 2 MCS is oriented in the reverse direction to this region within the pSPORT 1 plasmid. This modification eliminates any contribution from plasmid specific sequences, ensuring that any hybridization will be cDNA specific during subtractive hybridization.

Our protocol (figure 1) uses a biotinylated RNA driver produced from double-stranded pSPORT 2-cDNA plasmid and a single-stranded (ss) pSPORT 1-cDNA target plasmid DNA spiked with small amounts (1:2000 transformation ratio) of pBR322 plasmid DNA. Since this reporter plasmid should contain no complementary sequences to cDNA, it can be used to measure the subtraction efficiency. Alternatively, if the pBR322 plasmid is a concern, a parallel subtraction reaction can be performed without this reporter plasmid. The biotinylated driver RNA is hybridized to target, the common sequences subtracted and pBR322 and specific sequences are enriched. As another option, the biotinylated RNA driver can be produced from pSPORT 1-cDNA and ss DNA target can be prepared from pSPORT 2-cDNA. In this case, linearize the pSPORT 1 with *Not* I and transcribe with T7 RNA Polymerase to make driver RNA. The ss pSPORT 2-cDNA poly dT region must be blocked with poly $dA_{40\text{-}60}$. In addition to the above mentioned advantages of our system, it is easy to obtain miligram quantities of biotinylated driver RNA by *in vitro* transcription. Therefore, very concentrated driver can be added to the subtraction reaction, which will more easily drive the hybridization to completion. Finally, we utilize Rubenstein's (7) ds conversion procedure to improve the cloning efficiency of the unhybridized ss target DNA.

Applications

Screening Subtracted cDNA Libraries

After subtractive hybridization, unique or highly induced genes should be enriched. To identify these enriched genes, the cDNA inserts of randomly isolated clones from the subtracted library can be used to probe the target and subtracted cDNA libraries. If the relative level of a cDNA clone is higher in the subtracted library, most likely it will be a unique or highly induced gene.

Although random clone isolation is effective, it is tedious and cumbersome. A more efficient method takes advantage of the greater abundance of the enriched cDNA clones within the subtracted cDNA library. To identify all the enriched cDNAs, the subtracted cDNA library can be screened by colony hybridization using riboprobes made from this same subtracted cDNA library DNA. Clones containing enriched cDNAs should hybridize more strongly than subtracted clones that are present at reduced levels.

Normalized cDNA Libraries

Subtractive hybridization can be used to generate cDNA libraries that contain an equal representation of every expressed gene (17,18). These normalized (equalized) cDNA libraries, produced by self-hybridization at a high C_0t value (*i.e.*, 1,000), facilitate the isolation of rare cDNAs and reduce the redundancy encountered during the random sequencing of typical cDNA libraries. Additionally, an equalized driver library added to the normal driver population should increase the overall subtraction efficiency and remove more of the common sequences from the target library.

Developmental Stages

The pSPORT 1 and pSPORT 2 vectors are ideally suited for constructing developmental stage-specific cDNA libraries. For example, cDNAs from multiple stages during embryogenesis could be alternately cloned into these vectors. By hybridizing these libraries against one another, cDNAs representing the changes in gene expression should be rapidly enriched and identified.

Materials

cDNA Synthesis and Cloning Kits/Plasmid Vectors

Various cDNA synthesis and cloning kits are commmercially available (see product catalogues of Gibco/BRL, Stratagene, etc.). Most prokaryotic or eukaryotic plasmid vectors that contain an F1 origin of replication for the production of single-stranded DNA and RNA polymerase promoters (*ie.* T7, SP6, *etc.*) can be purchased from these same companies.

Eletrocompetent Cells

Certain bacterial strains (*i.e.* DH12S from Gibco/BRL) that contain the F episome may be used for the production of single stranded DNA for subtractive cDNA libraries.

Equipment

1. Electroporation pulse generator: An electroporation apparatus capable of generating a pulse length of 4 ms and a field strength of 16,600 V/cm (*i.e.* Cell-Porator Electroporation System I and Cell-Porator *E.coli* Electroporation System from Gibco/BRL).
2. Sterile electroporation chambers: Disposable chambers or cuvettes containing two electrodes separated by an average distance of 0.1 to 0.15 cm.
3. Bacterial incubator: Capable of maintaining a temperature of 37 °C and able to shake 500 ml to 2.8 L flasks at a setting of 275 rpm.

Advanced Preparations/Solutions

1. To ensure a low background of non-recombinant colonies, plasmid pSPORT 2 DNA (Gibco/BRL) can be digested with *Spe* I (Gibco/BRL; 2 units/µg plasmid) for 2 h at 50 °C, dephosphorylated with bacterial alkaline phosphatase (Gibco/BRL; 60 units/µg plasmid) for 1 h at 65 °C, digested with *Not* I (Gibco/BRL; 3 units/µg plasmid) for 2 h at 37 °C, followed by a *Sal* I (Gibco/BRL; 6 units/µg plasmid) digestion for 2 h at 37 °C. The pSPORT 2 DNA should be phenol:chloroform extracted 2 times and ethanol precipitated after each digestion procedure. Finally, the digested pSPORT 2 DNA can be further purified by agarose gel electrophoresis.
2. The *Not* I-oligo primer ($GCGGCCGCCCT_{15}$) must be synthesized.
3. 1X TE buffer: 10 mM Tris-HCl, pH 7.5, 1 mM EDTA. Autoclave and store at 4 °C.
4. SOB medium (without magnesium): Add 20 g of bacto-tryptone, 5 g of bacto-yeast extract, 0.584 g of NaCl, 0.186 g of KCl and distilled water to 800 ml. Mix components, adjust pH to 7.0 with NaOH, add distilled water to 1 L and autoclave. Store at 4 °C.
5. 2 M glucose: Add 36.04 g of glucose and distilled water to 100 ml. Mix components and filter sterilize. Store at 4 °C.
6. 2 M Mg^{++} stock: Add 20.33 g $MgCl_2•6H_20$ (Mallinckrodt), 24.65 g of $MgSO_4•7H_20$ (Mallinckrodt) and distilled water to 100 ml. Mix components and autoclave. Store at 4 °C.
7. SOC medium: Add 98 ml of SOB medium to 1 ml of sterile 2 M glucose and 1 ml of sterile 2 M Mg^{++} stock. Mix and store at 4 °C.
8. Glycerol solution (60% SOC, 40% glycerol)
9. DNase I solution (50 units/µl) for preparation of single-stranded DNA: Dissolve 2 mg of DNase I (1,500 units/mg, Sigma) in 60 µl of 20 mM Tris-HCl buffer, pH 7.5.
10. 40% (w/v) polyethylene glycol (PEG) solution: Add 200 g of PEG 4,000 (Fluka) and 250 ml of 5 M NaCl to a 1-liter beaker. Bring the volume to 500 ml with autoclaved water and stir until completely dissolved. Store at 4 °C.

12. Proteinase K solution (20 mg/ml): Dissolve proteinase K (100 mg/bottle, Gibco BRL) in 5 ml of 0.5 M EDTA, pH 8.0. Store at -20 °C.
13. Column buffer (maintain RNase free conditions): 10 mM Tris-HCl, pH 7.5, 1 mM EDTA and 100 mM NaCl in diethyl pyrocarbonate (DEPC) treated water. Autoclave and store at 4 °C.
14. Streptavidin binding buffer: 10 mM Tris-HCl, pH 7.5, 1 mM EDTA and 500 mM NaCl. Autoclave and store at 4 °C.
15. *Taq* DNA polymerase buffer: 20 mM Tris-HCl, pH 8.0, 50 mM KCl and 2.5 mM $MgCl_2$. Autoclave and store at 4 °C.
16. TEN buffer: 10 mM Tris-HCl, pH7.5, 1 mM EDTA and 200 mM NaCl. Autoclave and store at 4 °C.
17. 2X hybridization buffer: 80% formamide, 100 mM HEPES, pH 7.5, 2 mM EDTA and 0.2% SDS. Filter sterilize and store at 4 °C.

Subtractive Hybridization Methods

Protocol 1: Construction of Directional cDNA Libraries

Due to the abundant number of commercially available cDNA synthesis and cloning kits, the generation of double stranded cDNA from mRNA is now a less difficult process. The preferred cDNA kits are those that employ an $RNaseH^-$ reverse transcriptase to generate first strand cDNA. This enzyme produces higher yields of first strand cDNA and greater full-length cDNA synthesis than other reverse transcriptases (12,13). To optimize the cDNA cloning potential or colony forming units, various ligations containing different cDNA to vector mass ratios should be examined. Our protocol uses the Gibco BRL SuperScript Plasmid System for cDNA synthesis and plasmid cloning (14) to construct pSPORT 1-cDNA target (ampicillin resistance) and pSPORT 2-cDNA driver (ampicillin resistance) libraries.

1. Add the appropriate amount of cDNA (10, 20, or 40 ng) and plasmid vector (1 µl of pSPORT 1 or 2, *Not* I-*Sal* I-Cut at 50 ng/µl) to a sterile 1.5 ml microcentrifuge tube at room temperature. Add 4 µl of 5X T4 DNA ligase buffer and enough autoclaved water to bring the volume to 19 µl. Add 1 µl (1 unit) of T4 DNA ligase, and mix by pipetting.
2. Incubate this reaction for 3 h at room temperature.
3. Following incubation, add 5 µl (1 µg/ul) of yeast tRNA or 10 µg of glycogen and 12.5 µl of 7.5 M ammonium acetate (NH_4OAc) to the ligation reaction. Add 70 µl of absolute ethanol pre-cooled to -20 °C. Vortex the mixture thoroughly, and immediately centrifuge at room temperature for 20 min at 14,000 x g.
4. Remove the supernatant carefully, and overlay the pellet with 0.5 ml of 70% ethanol (-20 °C). Centrifuge for 2 min at 14,000 x g and remove the supernatant. Dry the ligated cDNA at 37 °C for 10 min to evaporate residual ethanol.
5. Add 5 µl of TE buffer to the dried pellet, vortex, and collect the contents of the tube by brief centrifugation. The DNA solution is now ready for electroporation.
6. Dispense 1 µl of DNA solution to each of 5 microcentrifuge tubes and place on ice. Always determine the transformation efficiency of the electrocompetent cells, by

including one control tube (1 µl=10 pg of plasmid DNA) for every set of electroporation reactions. This information can be used to trouble-shoot problems with the cDNA cloning.

7. Place the required number of sterile electroporation chambers (one per transformation) and frozen electrocompetent cells on ice.

8. After the cells are thawed (10 min at 4 °C), mix gently and transfer the appropriate amount of cells (24 µl per transformation using the Gibco/BRL Cell-Porator Electroporation System and 40 µl per transformation using the Bio-Rad Gene Pulser) to a microcentrifuge tube containing DNA. Using a micropipette, pipette this DNA-cell slurry gently to mix and place into an electroporation chamber between the electrode bosses.

9. Pulse this DNA-cell mixture once at 4 °C using the optimal conditions for the appropriate cell strain (a pulse length of approximately 4 ms and a field strength of 16,600 V/cm is used for DH12S).

10. After incubation of the electroporated cells for 1 h at 37 °C in 1 ml of S.O.C. media, plate appropriate dilutions of the cells on LB plates containing 100 µg/ml ampicillin and 0.01% X-gal. Incubate overnight at 37 °C.

11. Add 1 ml of glycerol solution to the remaining ~1 ml transformed cells, mix and freeze at -70 °C. The titer of these frozen transformed cells will remain unchanged for up to one year.

12. The following day count the number of viable colonies and determine the size of the cDNA library (total recombinants/electroporation) and the percentage of background blue colonies (vector containing no cDNA inserts). The size of the library is usually between 5 x 10^6 to 5 x 10^7 recombinants/ligation (For more information, see Note 1).

Protocol 2: Preparation of Single-Stranded DNA

Make a large scale ss plasmid DNA preparation from the pSPORT 1-cDNA target library.

1. Inoculate 200 ml LB broth containing 100 µg/ml ampicillin in a 1-liter flask with 1 x 10^6 pSPORT 1-cDNA transformants. Incubate the flask at 37 °C with shaking (275 rpm) for 3 h.

2. Add 200µl of M13KO7 helper phage (kanamycin resistance; >1 x 10^{11} pfu/ml) to the culture, and continue to incubate the culture for 2 h.

3. Add 1.5 ml of kanamycin (1%) to the cells for a final concentration of 75 µg/ml. Incubate the infected cells for an additional 22-24 h at 37 °C.

4. Centrifuge this culture in a GSA rotor at 10,000 rpm (16,270 x g) for 15 min at 4 °C.

5. Filter the supernatant through a 0.2 µm sterile Nalgene filter into an autoclaved GSA centrifuge bottle. Add 40 µl of DNase I (50 units/µl) and incubate at room temperature for 3 h. This step removes any residual ds pSPORT 1-cDNA plasmid contamination (For more information, refer to Note 2).

6. Transfer 100 ml of the supernatant to another GSA centrifuge bottle. Add 25 ml of 40% PEG 4000 solution to each of the centrifuge bottles containing the supernatant.

7. Vortex the mixture, incubate on ice for 1 h, and centrifuge in a GSA rotor at 10,000 rpm for 20 min at 4 °C.

8. Carefully discard the supernatant. To fully drain off the remaining solution from the pellets, place the bottles at an angle, with the pellet side facing up for 10-15 min. Remove the solution with a sterile Pasteur pipette.
9. Resuspend the pellets in 2 ml of TE buffer. Add 10 μl of proteinase K solution, 20 μl of 10% SDS and incubate this mixture at 45 °C for 1 h.
10. Transfer the digested mixture to 3 microcentrifuge tubes, and extract four times with equal volume of phenol:chloroform:isoamyl alcohol (25:24:1). Precipitate with 0.5 volume of 7.5 M NH_4OAc and 2.5 volume of ethanol. Store in dry ice for 10 min and centrifuge at 14,000 x g for 10 min at 4 °C. Remove the supernatant and wash the pellet with 0.5 ml of 70% ethanol. Dry the pellet and dissolve in 100 μl TE buffer.
11. To remove contaminating polysaccharide, freeze the solubilized DNA at -20 °C for 1 h and centrifuge in a microcentrifuge at 14,000g for 15 min at 4 °C.
12. Transfer the supernatant containing the ss plasmid DNA to a fresh tube, and discard the polysaccharide pellet. Store the ss DNA at 4 °C.
13. Determine the DNA concentration (OD_{260}) and the transformation efficiency. On average, 100 to 200 μg of ss DNA were obtained from 200 ml of cells.

Protocol 3: Preparation of Double-stranded DNA

Carry out a large preparation of double-stranded (ds) plasmid DNA from the pSPORT 2-cDNA driver library with the modifications listed below, using either the BRL NACS™37 Ion-Exchange Resin or, the method described by Sambrook *et al.* (15).

1. Inoculate 200 ml of LB broth containing 100 μg/ml ampicillin with 1 x 10^6 transformants from the pSPORT 2-cDNA driver glycerol stocks and incubate at 37 °C overnight.
2. After lysis but before NACS 37 purification or centrifugation in cesium chloride-ethidium bromide gradients, dissolve the plasmid DNA in 5 ml of TE and incubate at 65 °C for 10 min (to inactivate the *end* A protein from the DH12S cells).
3. After purification and dialysis, precipitate the plasmid DNA with ethanol and NH_4OAc as described above.
4. Resuspend the pellet in 200 μl of TE, add 1 μl of proteinase K solution and incubate the reaction mixture at 42 °C for 1 h.
5. Extract the plasmid DNA twice with phenol:chloroform:isoamyl alcohol (25:24:1), precipitate with ethanol and NH_4OAc, and dissolve in 100 μl of TE.

Protocol 4: Preparation of Biotinylated RNA Driver

Use the following protocol to generate biotinylated driver RNA from *Sal* I digested pSPORT 2-cDNA plasmid (prepared in Protocol 3). Our protocol utilizes the T7 RNA polymerase reaction from the GIBCO BRL Nonradioactive RNA Labeling System.

1. Digest a portion of the ds pSPORT 2-cDNA plasmid (20 μg) to completion with *Sal* I, extract with phenol:chloroform:isoamyl alcohol (25:24:1), precipitate with ethanol and NH_4OAc, and dissolve the DNA pellet in 20 μl of TE.
2. Incubate 10 μg of *Sal* I digested DNA in a 1 ml transcription reaction (40 mM

Tris-HCl, pH 8, 8 mM $MgCl_2$, 25 mM NaCl, 2 mM spermidine, 1 mM each ATP, UTP and GTP, 0.5 mM CTP, 2 mM biotin-14-CTP, 5 mM DTT and 500 units of T7 RNA polymerase) for 16 h at 37 °C in a 50 ml conical tube.
3. After 16 h, precipitate the reaction with ethanol and NH_4OAc, and dissolve the pellet in 2 ml of 20 mM Tris-HCl, pH 7.5.
4. Ethanol precipitate the Bio-RNA again, rinse the pellet with 75% ethanol and dissolve in 2 ml of 20 mM Tris-HCl (pH 7.5). Add $MgCl_2$ to a final concentration of 10 mM.
5. Add 200 units of RNase-free DNase I (this is different DNase I than used for the **Preparation of Single-Stranded DNA**) and incubate this reaction for 1 h at 37 °C (See Note 2, for more information).
6. After 1 h, transfer the reaction to a new 50 ml conical tube and add 80 μl of 0.25 M EDTA.
7. To inactivate the DNase I, incubate the RNA mixture for 10 min at 65 °C and precipitate with ethanol and NH_4OAc.
8. Completely dissolve the biotinylated RNA in 200 μl of column buffer (**Note: The RNA may need to be heated at 65 °C for 5 min**). To remove free biotin-14-CTP from the biotinylated RNA, load the RNA mixture onto a Sephadex G-50 column (1.0 x 18 cm), pre-equilibrated with column buffer.
9. Pool the RNA fractions, precipitate with ethanol and NH_4OAc, and dissolve the pellet in 50 μl of TE. Dilute 1 μl of RNA in 500 μl TE and determine the concentration (OD_{260}). Typically, each transcription reaction yields 250 to 300 μg of RNA after purification.
10. To determine the amount of ds DNA remaining in the purified RNA, mix 20 μg of biotinylated RNA with 50 μl of streptavidin binding buffer. To this mixture, add 25 μg of streptavidin that has been resuspended in streptavidin binding buffer at 5 μg/μl, vortex and incubate at room temperature for 5 min.
11. Extract with phenol:chloroform:isoamyl alcohol two times. Add 1 μl of glycogen (10 μg/μl), 25 μl of 7.5 M NH_4OAc and 190 μl of pre-cooled (-20 °C) absolute ethanol to the aqueous phase. Vortex the mixture thoroughly, and immediately centrifuge at room temperature for 20 min at 14,000 x g.
12. Dissolve the recovered DNA in 5 μl of TE and electroporate 2 μl directly into ElectroMAX DH12S cells (see Protocol 1).
13. Spread 100-500 μl of the 1 ml culture onto LB-ampicillin plates. If the ds plasmid contamination is <500 colonies per 100 μg RNA, use the biotinylated RNA in the subtractive hybridization. If >500 colonies, take the RNA back through steps 4-13.

Protocol 5: Pre-Hybridization Blockage of ss pSPORT 1-cDNA Poly dA Regions

To prevent any nonspecific hybridization between the poly dA region of the ss pSPORT 1-cDNA target and the poly U region of the biotinylated RNA driver, the poly dA region of the cDNA must be pre-blocked.

1. Dissolve 20 μg of ss pSPORT 1-cDNA and 4 μg of *Not* I-oligo primer ($GCGGCCGCCCT_{15}$) in 60 μl of *Taq* DNA polymerase buffer and heat at 90 °C for 2 minutes.
2. Transfer the mixture to a 55 °C water bath and incubate for 30 min to anneal the

primer to the ss DNA.
3. Add 60 µl of prewarmed (55 ºC) *Taq* DNA polymerase mixture containing 600 µM dTTP and 20 units of *Taq* DNA polymerase in *Taq* buffer to the preannealed ss DNA.
4. Transfer the reaction mixture to a 70 ºC water bath and incubate for 20 min.
5. After 20 min, add 6 µl of 0.25 M EDTA and extract the reaction mixture once with an equal volume of phenol:chloroform:isoamyl alcohol (25:24:1).
6. Back extract the organic phase with 30 µl of TE, pool the aqueous phases, ethanol precipitate and dissolve the DNA pellet in 20 µl of TEN buffer.

Protocol 6: Subtractive Hybridization

Use the following protocol to perform a subtractive hybridization to a C_0t value of 1000 (For information about C_0t values, see Note 3). The hybridization reaction consists of 600 ng of target DNA (poly dA blocked ss pSPORT 1-cDNA plasmids spiked with small amounts [5 picograms] of pBR322 plasmid [ampicillin and tetracycline resistance] DNA) and 86 µg of biotinylated RNA driver. The pBR322 reporter plasmid is necessary to quantitate the subtraction and repair efficiencies (For more information about subtraction and repair efficiencies, see Note 4).

1. Add the driver RNA to 25 µl of 2X hybridization buffer in a 5 ml Falcon tube.
2. Heat this driver RNA mixture at 65 ºC for 10 min, quick chill on ice and add the target ss DNA (600 ng) to the RNA mixture.
3. Add 2 µl of 5 M NaCl to the RNA-DNA mixture and adjust the total volume to 45 µl with DEPC treated water.
4. To increase the rate and extent of the subtractive hybridization (16), incubate the hybridization mixture at 42 °C with shaking (200 rpm) in a bacterial incubator for 48 h. (The control experiment contains all components except the biotinylated RNA.)
5. After hybridization, transfer the mixture to an eppendorf tube, add 25 µg of streptavidin that has been resuspended in streptavidin binding buffer at 5 µg/µl, vortex and incubate the mixture at room temperature for 5 min.
6. Extract with phenol:chloroform:isoamyl alcohol (25:24:1), back extract the organic phase with 50 µl of streptavidin binding buffer and pool the aqueous phases. The streptavidin-biotinylated RNA driver-DNA complexes enter the organic phase.
7. Repeat the streptavidin binding and extraction two more times and ethanol precipitate the unhybridized target DNA with 5 µg of GIBCO BRL yeast tRNA or 10 µg of glycogen as a carrier. Store in dry ice for 10 min, and centrifuge for 10 min at 4 ºC.
8. Resuspend the pellet in 10 µl of TE.
9. Place 25 ml of TE buffer into a plastic petri dish.
10. Float a 0.025 µm Millipore filter on top of the TE buffer.
11. After 10 min, place the target DNA (10 µl) onto the filter and dialyze for 30 min at room temperature. This step removes any impurities (*i.e.*, SDS) that might affect subsequent steps.
12. Collect the dialyzed sample and electroporate an aliquot (2 µl) directly into ElectroMAX DH12s cells (see **Protocol 1: Construction of directional cDNA libraries**) or, to increase the number of transformants, repair using *Taq* polymerase

(see **Protocol 7: Repair of Single-Stranded DNA**) and then electroporate into ElectroMAX cells.

13. After 1h incubation at 37 °C, plate the cells differentially onto ampicillin and tetracycline plates to determine the enrichment of the pBR322 plasmid (before repair) or to measure the effectiveness of the repair step (see Note 4 and Table 1). If pBR322 is not added, do not plate onto tetracycline plates.

Table 1. The Enrichment of the Tc^R Gene in Subtracted cDNA Library

	Reaction	Ap^R	Tc^R	Tc^R/Ap^R	Enrichment of Tc^R (fold)
Before repair	Control	3.9×10^5	65	1/6000	-
	Sample	7.8×10^3	79	1/99	61
Repair	Control	3.0×10^7	86	$1/3.5 \times 10^5$	-
	Sample	1.5×10^5	79	1/1898	184

The Sample and Control refer to the subtracted (C_0t=1,000) and control (C_0t=0) cDNA library. The abbreviations represent: Ap^R, ampicillin resistant colonies; Tc^R, tetracycline resistant colonies. The enrichment of Tc^R gene was calculated by dividing the ratio of Tc^R/Ap^R of the sample by the corresponding ratio of the control.

Protocol 7: Repair of Single-Stranded DNA

After subtraction, the remaining ss target DNA represents a very small percentage of the starting material (1-10 ng). This ss DNA transforms electrocompetent cells less efficiently than ds DNA. By converting the ss target DNA to ds DNA (7,10,11), the target DNA can better compete in the subsequent transformation step effectively reducing the apparent levels of ds plasmid DNA contamination and insertless ss DNA (see Notes 1 and 2). This method takes advantage of the partially repaired ss pSPORT 1-cDNA and uses the procedure as found in **Protocol 5: Pre-Hybridization Blockage of ss pSPORT 1-cDNA Poly dA Regions** with the following modifications:

1. Perform the reaction in a 30 µl volume containing the subtracted ss pSPORT 1-cDNA product, 50 ng of *Not* I-oligo primer, *Taq* DNA polymerase buffer, 300 µM dNTP mix (dATP, dTTP, dGTP and dCTP) and 1 unit of *Taq* DNA polymerase (In addition, the 90 °C denaturation step should be eliminated).
2. After repair, extract the mixture once with phenol:chloroform:isoamyl alcohol (25:24:1), wash the organic phase with 30 µl of TE, pool the aqueous phases and ethanol precipitate with carrier.
3. Rinse the pellet with 100 µl of 75% ethanol, resuspend in 10 µl of TE and electroporate an aliquot (2 µl) into ElectroMAX cells.
4. After 1h incubation at 37 °C, plate the cells differentially onto ampicillin and tetracycline plates to measure the effectiveness of the repair step (see Note 4). If pBR322 is not added, do not plate onto tetracycline plates.
5. The next day, after calculating the subtraction and repair efficiencies, electroporate enough repaired DNA to yield 1-2 x 10^5 clones.
6. After plating one half of this electroporation onto ampicillin plates (2,000-5,000/100 mm plate), grow the remaining clones overnight as found in **Protocol 3: Preparation of Double-stranded DNA**.

7. Isolate the ds DNA and refer to the Applications section: **Screening Subtracted cDNA Libraries**.

Notes

1. Subtractive hybridization will enrich for any target nucleic acid sequence that does not have a complement in the driver population. Therefore, any ss pSPORT 1 target plasmid DNA lacking cDNA inserts or clones containing only poly dA regions will be enriched after subtraction. During cDNA synthesis and cloning, care must be taken to remove small inserts (*i.e.*, stringent cDNA size fractionation) and to reduce the vector background (*i.e.*, different cDNA to vector mass ratios should be examined). Nevertheless, the apparent level of vector background will be minimized following the repair step.

2. Double-stranded plasmid contamination of the ss target plasmid DNA or the biotinylated RNA driver will be enriched artificially during subtractive hybridization. This contamination must be removed by DNase I treatments of the phagemid preparation and the RNA driver. The efficiency of the DNase I digestion can be examined by sequencing the ss target DNA with the M13 and T7 sequencing primers. After DNase I treatment, sequence information should only be generated with the M13 primer. Additionally, the effectiveness of the DNase I digestion of the ds DNA template in the RNA driver can be measured by electroporating a portion of this driver into electrocompetent bacterial cells before and after this treatment (see **Protocol 4: Preparation of Biotinylated RNA Driver**). However, the repair step should reduce the transformation contribution of any remaining ds DNA contamination.

3. C_0t is the product of the starting concentration of nucleic acid (moles of nucleotide per liter, C_0) and time (secs, t). The unit of C_0t is normally expressed as mol x s x L^{-1} or M x s. Generally, a nucleic acid population at 90 µg/ml that is hybridized for 1h is approximately equal to a C_0t of 1. C_0t values increase as greater RNA driver concentrations are added to a constant amount of DNA target or with increasing hybridization time. As C_0t increases, more common sequences should be hybridized and removed and unique or highly induced sequences should become more enriched. Usually, high C_0t values are required to effectively remove the majority of common sequences between two closely related DNA populations.

4. The pBR322 reporter plasmid should contain no sequence complementarity to cDNA inserts. Although double-stranded, this plasmid can be used to approximate the enrichment expected for a real (ss DNA) gene. To quantitate the subtraction efficiency a small amount of reporter plasmid is added to the ss target DNA library. A portion of this library is hybridized to driver RNA (sample reaction) and another portion is handled identically as the sample without adding driver RNA (control reaction). The reporter plasmid should remain unchanged and serve as an internal control for measuring the overall reduction of the ssDNA in the sample, whereas the complexity of the ssDNA in the control reaction should not change. The enrichment of the reporter plasmid (Table 1) is calculated by dividing the ratio of tetracycline resistant colonies (Tc^R)/ampicillin resistant colonies (Ap^R) of the sample by the corresponding ratio of the control reaction.

Generally, at a C_0t value of 1,000, the pBR322 plasmid is enriched 34-70 fold before repair (10, 11). Additionally, the subtractive hybridization efficiency can be measured by monitoring the reduction of a common abundant cDNA (*i.e.* actin). At a C_0t value of 1,000, the actin clones in the subtracted cDNA library should be reduced >99.5% (10, 11). After subtraction, if the pBR322 enrichment is lower or actin clone number is higher than the expected values, add more RNA driver or hybridize longer than 48 h.

To determine the efficiency of the repair step, the ratio of Tc^R/Ap^R colonies before and after repair should be compared. After repair, a 20-100 fold increase in the number of Ap^R colonies is typically observed.

References

1. Britten, R.J. and Davidson, E.H. 1985. Hybridization Strategy. In: Nucleic Acid Hybridization. IRL Press, Oxford and Washington D.C.
2. Sive, H.L. and John, T.St. 1988. A simple subtractive hybridization technique employing photoactivatable biotin and phenol extraction. Nucl. Acids Res. 16: 10937.
3. Rothstein, J.L., Johnson, D., Jessee, J., Skowronski, J., Deloia, J.A., Solter, D. and Knowles, B.B. 1993. Construction of primary and subtracted cDNA libraries from early embryos. In: Methods in Enzymology. Academic Press, Inc. Vol. 225: P. 587-611.
4. Lopez-Fernandez, L.A. and Mazo, J.D. 1993. Construction of subtractive cDNA libraries from limited amounts of mRNA and multiple cycles of subtraction. BioTechniques. 15: 654-658.
5. Sharma, P., Lonneborg, A. and Stougaard, P. 1993. PCR-based construction of subtractive cDNA library using magnetic beads. BioTechniques. 15: 610-611.
6. Duguid, J.R., Rohwer, R.G., and Seed, B. 1988. Isolation of cDNAs of scrapie-modulated RNAs by subtractive hybridization of a cDNA library. Proc. Natl. Acad. Sci. USA. 85: 5738-5742.
7. Rubenstein, J.L.R., Brice, A.E.J., Ciaranello, R.D., Denney, D., Porteus, M.H. and Usdin, T.B. 1990. Subtractive hybridization system using single-stranded phagemids with directional inserts. Nucl. Acids Res. 18: 4833-4842.
8. Swaroop, A., Xu, J., Agarwal, N. and Weissman, S.M. 1991. A simple and efficient cDNA library subtraction procedure: isolation of human retina-specific cDNA clones. Nucl. Acids Res. 19: 1954.
9. Swaroop, A. 1993. Construction of directional cDNA libraries from human retinal tissue/cells and their enrichment for specific genes using an efficient subtraction procedure. In: Methods in Neurosciences. Academic Press, Inc. Vol. 15: p. 285-300.
10. Gruber, C.E., Li, W.-B., Lin, J.-J. and D'Alessio, J.M. 1993. Subtractive cDNA hybridization using the multifunctional plasmid vector pSPORT 2. Focus. 15: 59-65.
11. Li, W.-B., Gruber, C.E., Lin, J.-J., Lim, R., D'Alessio, J.M. and Jessee, J.A. 1994. The isolation of differentially expressed genes in fibroblast growth factor stimulated BC3H1 cells by subtractive hybridization. BioTechniques. 16: 722-729.

12. Kotewicz, M.L., Sampson, C.M., D'Alessio, J.M., and Gerard, G.F. 1988. Isolation of cloned Moloney murine leukemia virus reverse transcriptase lacking ribonuclease H activity. Nucl. Acids Res. 16: 265-277.
13. Gerard, G.F., D'Alessio, J.M., and Kotewicz, M.L. 1989. cDNA synthesis by cloned Moloney murine leukemia virus reverse transcriptase lacking RNase H activity. Focus. 11: 66-69.
14. D'Alessio, J.M., Gruber, C.E., Cain, C. and Noon, M.C. 1990. Construction of directional cDNA libraries using the SuperScript plasmid system. Focus. 12: 47-48.
15. Sambrook, J., Fritsch, E.F. and Maniatis, T. 1989. Molecular Cloning: A Laboratory Manual. Cold Spring Harbor Laboratory Press, Cold Spring Harbor, New York.
16. Ness, J.V. and Hahn, W.E. 1982. Physical parameters affecting the rate and completion of RNA driven hybridization of DNA: new measurements relevant to quantitation based on kinetics. Nucl. Acids Res. 10: 8061-8077.
17. Ko, M.S.H. 1990. An equalized cDNA library by the reassociation of short double-stranded cDNAs. Nucl. Acids Res. 18: 5705-5711.
18. Patanjali, S.R., Parimoo, S. and Weissman, S.M. 1991. Construction of a uniform-abundance (normalized) cDNA library. Proc. Natl. Acad. Sci. USA. 88: 1943-1947.

From: *Molecular Biology: Current Innovations and Future Trends*.
ISBN 1-898486-01-8 ©1995 Horizon Scientific Press, Wymondham, U.K.

8

OLIGORIBONUCLEOTIDES: THEORY AND SYNTHESIS

Ravi Vinayak

Abstract

Chemical synthesis of biologically active oligoribonucleotides (RNA) fills a unique role for precise nucleotide substitution and large scale requirements for physical studies, antisense and ribozyme investigations. The growing number of applications for synthetic RNA have generated a demand for more effective methods for synthesis and post-synthesis protocols. In this chapter, chemical synthesis of oligoribonucleotides based on solid-phase chemistry using cyanoethyl phosphoramidites is detailed. Improved post-synthesis protocols are presented that assure isolation of high-quality synthetic RNA. Also, reliable techniques of analysis and purification, based on reverse phase and anion exchange HPLC are described.

Introduction

Interest in ribozymes, catalytic RNA and antisense RNA has propagated the demand for chemically synthesized RNA (1-6). Catalytic RNAs have the intrinsic ability to break and form covalent bonds. These molecules catalyze the cleavage of RNA substrates intra- and intermolecularly. Catalytic activity demands high specificity and turnover in the cleavage reaction (7, 8). It has been demonstrated that a ribozymal activity is essential for the formation of peptide bonds during protein synthesis (9-11). The targeting of ribozymes to cleave viral RNAs *in vitro* and exploitation of the ribozyme catalytic center for cleavage of a specific RNA transcript are emerging as potential therapeutic applications (12-14). Ribozymes have become important tools for inhibiting gene expression in many biological systems (15). Efforts are also being directed towards developing diagnostic and therapeutic reagents based on oligoribonucleotides (1-3). There has also been an intense interest in chemical synthesis of oligoribonucleotides for physical and structural studies (16-18). These and other new applications for synthetic oligoribonucleotides have escalated the demand for more effective methods for large scale production of biologically active RNA (19).

The biological activity of chemically synthesized RNA, comparable to that of RNA derived by transcription methods, is contingent upon efficient synthesis and purification systems (20-22). Solid-phase chemistry currently provides the most effective means for the scale-up of RNA synthesis. The chemical synthesis of RNA, like DNA is carried out in the 3'- to 5'-direction to take advantage of the high chemical reactivity of the 5'-hydroxyl group, in contrast to enzymatic biosynthesis which proceeds in the reverse direction. Progress in the methods for automated solid-support RNA synthesis has lagged behind DNA synthesis primarily due to problems of 2'-hydroxyl protection, slower internucleotide coupling kinetics, and the fragility of RNA to hydrolytic and enzymatic degradation (23, 24).

RNA copies from DNA templates can be made by using either SP6 or T7 RNA polymerases (25, 26). A general method for the preparation of isotopically enriched RNAs of defined sequence involving *in vitro* transcription has been published (27, 28). Using this method, milligram quantities of RNAs are enzymatically synthesized using isotopically labeled NTPs . However these methods require that a DNA substrate be available for transcription or that one be synthesized and/or constructed. The combined use of T4 RNA ligase and polynucleotide kinase allows the joining of small building blocks to make long RNA molecules (29). However, not all sequences can be synthesized efficiently using enzymatic procedures.

Today, almost all synthetic oligonucleotides are prepared by solid-phase phosphoramidite chemistry (30). Such methods proceed by stepwise addition of protected nucleoside phosphoramidite monomers to a growing oligonucleotide chain attached to a solid support. These chemical reactions are driven to completion by large excesses of solution reagents relative to the polymer-bound nucleoside. As the growing oligonucleotide chain is attached to the solid support, all excess reagents are washed away, eliminating time-consuming purification after each monomer addition.

The chemical synthesis of oligoribonucleotides is of increasing importance as it enables production of the large quantities of RNA needed for studies in biochemistry

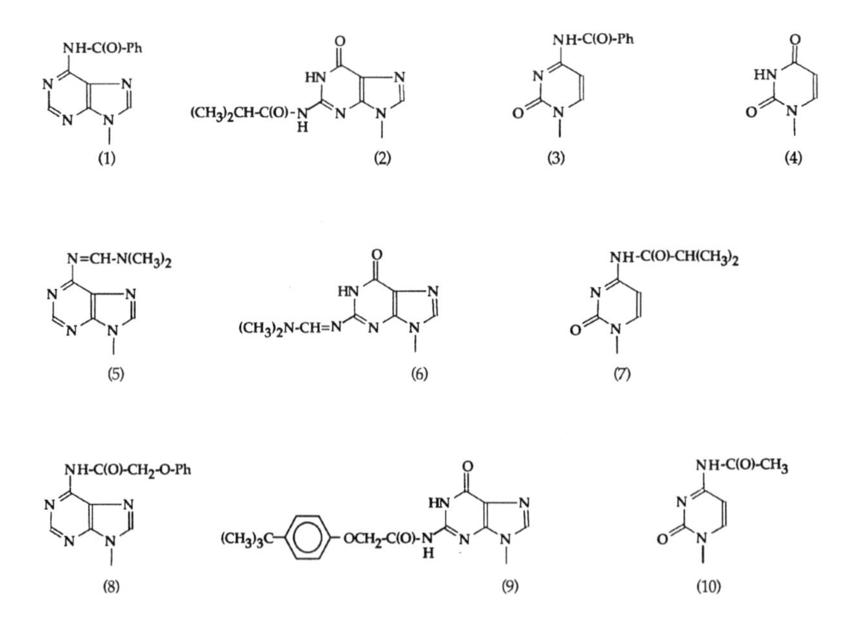

Figure 1. Standard {(1) A-Bz, (2) G-iBu, (3) C-Bz, (4) U} and fast deprotecting {(5) A-dmf, (6) G-dmf, (7) C-iBu, (8) A-PAc, (9) G-t-BPA, (10) C-Ac} groups used in solid phase RNA synthesis.

and processing of RNA, as well as for NMR, X-ray crystallography and therapeutic ribozyme investigations. Also, it allows for site-specific modifications such as deoxynucleotides (31, 32), phosphorothioates (33, 34), unnatural and modified nucleotides (35, 36) and branched oligoribonucleotides (37). The synthesis of triplex-forming circular RNAs has also been reported with potential applications in nucleic acid recognition (38).

Problems associated with automated RNA synthesis such as low purity of ribonucleoside monomers, low coupling efficiency, low biological activity of the final product etc., have been overcome in recent years by improving the protecting groups and with the development of new techniques of purification. RNA monomers and supports are currently readily available from a number of commercial sources that assure high quality synthetic RNA.

Several review articles on the chemical synthesis of RNA have dealt with protecting group strategies and synthesis and purification methodologies (20, 23, 25, 39, 40). This chapter will briefly review the most common chemistries used for RNA synthesis, and reliable methods of isolation, analysis and purification.

Protecting Groups

Base Protecting Groups

The traditionally used protecting groups for exocyclic amine functionality in ribonucleosides are benzoyl (Bz) for adenine (A) and cytosine (C) and isobutyryl (iBu) for guanine (G) (Figure 1; 41, 42). Phosphoramidite nucleoside monomers with these base protecting groups are available commercially. Deprotection of these base-protecting groups requires 16-24 hours at 55 °C in an ammonia solution. This extended exposure to ammonia has the potential to cause problems like hydrolysis of the 2'-protecting group (generally alkylsilyl) and internucleotidic cleavage resulting in low biological activity of the chemically synthesized RNA (compared with transcribed RNA).

These problems have been overcome with the use of base-labile protecting groups like phenoxyacetyl (PAc) or *t*-butylphenoxyacetyl (*t*-BPA) for adenine and guanine and acetyl (Ac) for cytosine (43-45). The mild deprotection conditions (less than 1 hour at 55 °C) minimize premature desilylation during the post-synthetic work-up of RNA thereby preserving the integrity of RNA.

Use of (dimethylamino)methylene protecting groups (Figure 2) for RNA synthesis has also been reported (46). In this work, the exocyclic amine groups of adenine and guanine have been replaced with (dimethylamino)methylene while the amino group of cytosine is protected as isobutyryl. Complete deprotection of these groups has been achieved within 2-3 hours at 55 °C in an ethanolic ammonia solution. Synthesis of 50

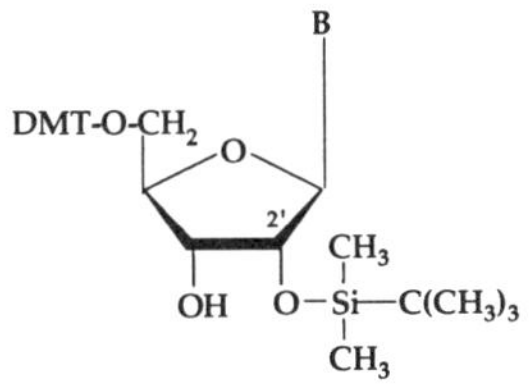

Figure 2. t-Butyl-dimethylsilyl (TBDMS) group (for 2'-OH protection) used in commercially available RNA phosphoramidites and supports.

nucleotide long ribozymes with full catalytic activity has been successful using these RNA phosphoramidites (46). Chemical synthesis of biologically active RNA using (dimethylamino)methylene protecting groups for purine nucleosides based on H-phosphonate chemistry has also been reported (47).

2'-OH Protecting Groups

A wide array of 2'-hydroxyl (2'-OH) protecting groups has been reviewed in the literature (20, 23, 25, 40, 48). Some of these, however, suffer from drawbacks and limitations resulting in poor oligoribonucleotide synthesis thereby limiting their commercialization.

The most commonly used protecting group for 2'-OH in RNA synthesis is the *t*-butyldimethylsilyl (TBDMS) group (Figure 2; 49, 50). Introduction of this group has made a significant impact on the practicability of RNA synthesis. The TBDMS group is stable to acidic and basic conditions used during automated synthesis and can be removed under mild conditions using either tetrabutylammonium fluoride (Bu_4NF) or triethylamine trihydrofluoride $\{Et_3N(HF)_3\}$.

Reese and coworkers have described the use of an acetal protecting group: 1-(2-fluorophenyl)-4-methoxypiperidin-4-yl (Fpmp) for 2'-OH protection (51). This modified acetal is stable during non-aqueous acidic treatment to remove 5'-O-DMT group, but is easily removed under aqueous conditions at low pH. Successful synthesis of oligoribonucleotides up to 37 bases in length has been reported using this approach (51, 52). Phosphoramidites bearing either 2'-O-TBDMS or 2'-O-Fpmp groups are available commercially.

5'-OH and Phosphate Protecting Group

The 4,4'-dimethoxytrityl (DMT) group for 5'-OH and ß-cyanoethyl group for phosphate are the most popular and widely used protecting groups in oligonucleotide synthesis. The DMT group has been widely accepted in oligonucleotide synthesis because of the ease of monitoring of the coupling efficiency as measured by the trityl cation release (53). ß-Cyanoethyl group for phosphate protection in the phosphite triester approach significantly simplifies and reduces the time necessary for deprotection and work-up of the final oligonucleotide product on polymer support (54).

Automated RNA Synthesis

The Solid-Support

The solid phase approach for RNA synthesis requires a solid support which is attached to the ribonucleoside to act as the 3' end of an oligonucleotide chain. During cleavage of the oligoribonucleotide from the support, the nucleoside attached to the support becomes the 3' residue. The nucleoside is generally attached to the solid support through a succinyl linker (55). The nucleoside succinate, synthesized by coupling the nucleoside with succinic anhydride, is then coupled onto controlled pore glass (CPG) beads (55,

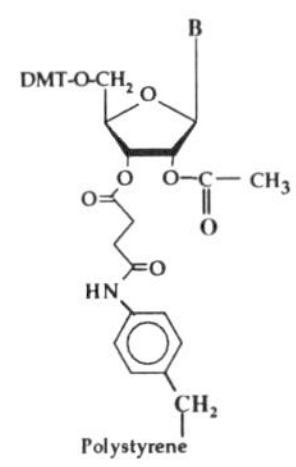

Figure 3. Polystyrene support for RNA (1000 Å pore size, 50-70 μm particle size).

56). The nucleoside succinates have also been loaded onto a polystyrene based support (46, 57, 58) which has low affinity for water and wets more thoroughly with organic solvents than CPG. This results in efficient removal of water by anhydrous acetonitrile during automated synthesis. Non-swelling highly cross-linked polystyrene beads possess other features like rapid reaction kinetics and better mechanical stability (57). The polystyrene support has been shown to facilitate efficient synthesis of RNA demonstrated through synthesis of ribozymes (46). The 2'-O-acetyl protected ribonucleosides have also been loaded onto polystyrene support (Figure 3). Acetyl is preferred as the 2'-OH protecting group for RNA supports, instead of the sterically bulky 2'-O-*t*-butyldimethyl silyl group since the latter hinders the complete cleavage of the RNA from the solid support (59).

Synthesis of an RNA Oligonucleotide on an Automated Synthesizer

The cycle of assembly of an oligoribonucleotide proceeds as shown in figure 4. In the first step of the synthesis, the 5'-O-DMT group on the nucleoside attached to the solid support (3' end of the oligoribonucleotide) is detritylated with trichloroacetic acid/ dichloromethane (step I). The RNA phosphoramidite (the chemical synthesis of a protected ribonucleoside phosphoramidite monomer has been described in reference 21) and tetrazole are then delivered to the column. The phosphoramidite is activated by tetrazole and coupling to the free 5'-OH group of the 3' bound nucleoside takes place (step II). The coupling takes about 10 minutes. This long coupling time for RNA may be attributed to the presence of the sterically bulky 2'-O-TBDMS group adjacent to the reactive 3'-O-phosphoramidite group. Use of 5-(4-nitrophenyl)-1*H*-tetrazole as an alternate activator in order to reduce the coupling time has been reported (52). The use of 5-ethylthio-1*H*-tetrazole as an activator has been recently reported also. This has been found to give enhanced coupling efficiencies compared with tetrazole and the differences are especially pronounced at higher synthesis scales and lower phosphoramidite excess (60). The greater acidity and solubility of 5-ethylthio-1*H*-tetrazole perhaps accounts for the superior performance (61). This is followed by capping of the unreacted nucleotide with a mixture of 1-methyl-imidazole/ tetrahydrofuran (THF) and acetic anhydride/lutidine/THF (step III). The oxidation of the P(III) species to the phosphate is then effected with iodine/pyridine/water (step IV). *t*-Butyl hydroperoxide is a useful alternative to iodine for oxidation of the internucleotide phosphite linkage during oligoribonucleotide chain elongation (61). Further repetitions of this cycle using either DNA or RNA phosphoramidites yields the oligonucleotide of desired length and sequence.

Deprotection of the product is a three step process (Figure 5). It is essential for the protecting groups to be removed in the correct order. The first step involves cleaving the oligoribonucleotide bound to the solid support with a 3:1 mixture (v/v) of 30% aqueous ammonium hydroxide : absolute ethanol (62; step V) which also removes the 2-cyanoethyl (phosphate protecting) group. In the next step this solution is heated at 55 °C to remove the base protecting groups (step VI). The inclusion of ethanol is necessary to adequately solubilize the more hydrophobic 2'-O-silyl-oligonucleotide and to minimize desilylation that can occur with concentrated ammonium hydroxide. In the final step, the 2'-O-silyl protecting groups are removed (step VII) to yield the fully deprotected oligoribonucleotide. This is then desalted (step VIII) to avoid distortion during analysis and/or purification. Desalting procedures generally remove inorganic salts, traces of organic compounds, and other low molecular weight impurities and some failure sequences (shorter sequences that failed to couple). This can be accomplished in a number of different ways (20-22). Desalted oligoribonucleotides are then purified. The use of a good purification procedure for the synthetic product is of vital importance for the success of the solid-phase method. Some of the purification techniques are described in the Protocols section.

Reagents Required for RNA Synthesis on an Automated DNA/RNA Synthesizer

The reagents and solvents used in solid-phase RNA synthesis should be of the highest purity. Care must be exercised at every step to ensure good quality of the final product. It is essential to use fresh reagents, anhydrous acetonitrile (for dilution of the phosphoramidites), gloves, and sterile material for high quality synthetic RNA.

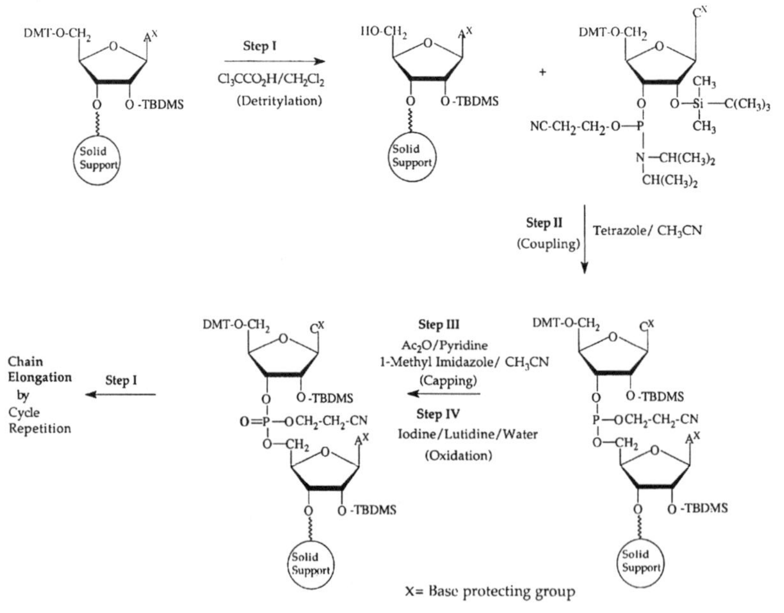

Figure 4. Cycle of assembly of an oligoribonucleotide on an automated DNA/RNA synthesizer.

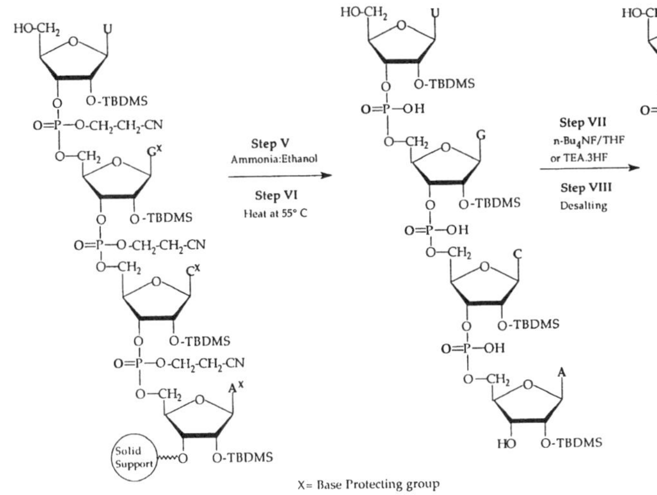

Step V
Ammonia:Ethanol
Step VI
Heat at 55° C
Step VII
n-Bu4NF/THF
or TEA.3HF
Step VIII
Desalting
Solid Support
X= Base Protecting group

Figure 5. Cleavage and deprotection of the oligoribonucleotide.

We have routinely synthesized oligoribonucleotides (Figure 6, see also reference 88) on Applied Biosystems DNA/RNA synthesizers (Models 380B, 391 EP and 392/394) using Applied Biosystems ancillary reagents according to recommended cycles, procedures and protocols (62).

Ribonucleoside Phosphoramidites (Bearing 2'-TBDMS Group)

1. A-RNA phosphoramidite
2. G-RNA phosphoramidite
3. C-RNA phosphoramidite
4. U-RNA phosphoramidite

1
2
3
24.0
22.0
20.0
18.0
16.0
14.0
12.0
10.0

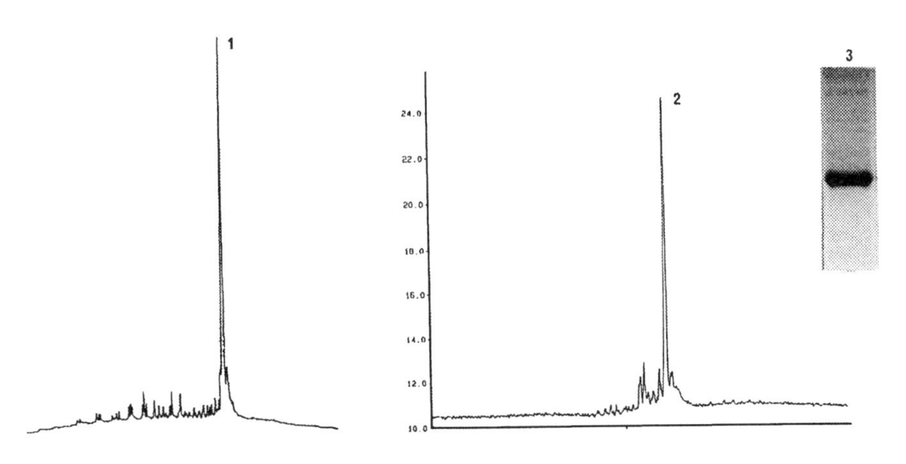

Figure 6. A 21 mer RNA { 5'> CUC AAU AAA GCU UGC CUU GAG <3'} (88) obtained after TBAF desilylation and G-25 Sephadex desalting. 1, anion exchange HPLC (at 50 °C); 2, MicroGel CE; 3, PAGE analysis.

RNA phosphoramidites are diluted with anhydrous acetonitrile to a concentration of 0.1 M and should be used immediately for best results. The phosphoramidite reagents are generally stable up to 3 days on the instrument after dilution. Exposing the reagents or solutions to moisture results in degradation of the phosphoramidites, significantly affecting synthesis performance. Dilution volumes for each phosphoramidite is different because of difference in the molecular weights.

Derivatized Nucleoside Functionalized Supports (Either CPG or Polystyrene Based)

1. A, 2. G, 3. C and 4. U.

Ancillary Reagents

These are the same as used in DNA synthesis and are used at the same bottle positions. The consumption of each reagent per cycle is virtually the same for both DNA and RNA.

1. Tetrazole/ acetonitrile
2. Trichloroacetic acid/ dichloromethane
3. Acetic anhydride/ lutidine/ THF
4. 1-Methyl-imidazole/ THF
5. Iodine/ water/ pyridine
6. HPLC grade acetonitrile (bottle position 18, on any Applied Biosystems DNA/ RNA synthesizer)
7. 30% Ammonium hydroxide/ Absolute ethanol (3:1 mixture).

Synthesis of an oligoribonucleotide (DMT-Off or DMT-On) on an automated synthesizer can be carried out using the appropriate scale RNA cycle. Syntheses are generally conducted DMT-Off, as there is a possibility of partial loss of the 5'-O-DMT group during desilylation. Purification of RNA oligonucleotides bearing the 5'-O-DMT group requires deprotection of the 2'-TBDMS group prior to removal of the 5'-O-DMT group. Exposing the fully deprotected RNA to acidic conditions (detritylation) allows the possibility of phosphate migration. Isolation and purification of oligoribonucleotides bearing the 5'-O-DMT group can be carried out by the butanol precipitation procedure (63).

Phosphorothioate oligoribonucleotides can be synthesized by substituting the oxidizing reagent (iodine/water/pyridine) with "Beaucage reagent"(64) or bis(O,O-diisopropoxy phosphinothioyl)disulfide (S-tetra, 65) or TETD (tetraethylthiuram-disulfide/acetonitrile) (66).

Cleavage and Deprotection of the Oligoribonucleotide

The fully protected oligoribonucleotide (after automated synthesis) is cleaved from the support on the synthesizer (using pre-programmed or user-defined end procedures) or manually (by the double syringe method) with a 3:1 mixture (v/v) of 30% aqueous ammonium hydroxide : absolute ethanol (62). This solution is then heated at 55 °C for 4 to 5 hours to remove the base protecting groups.

As an alternative to this routinely used aqueous ammonia : ethanol deprotection, use of anhydrous ammonia in ethanol has been reported with no degradation in the final product (41). However, the volatile nature of this reagent precludes its use on an automated instrument.

The resulting 2'-O-silyl-oligonucleotides are relatively stable compounds. They are resistant to ribonucleases and may be stored for extended times as dry pellet or in cold (-20 °C), aqueous, neutral solutions.

Removal of 2'-O-Silyl Group

The 2'-O-silyl-oligoribonucleotides are generally desilylated with tetrabutylammonium fluoride (TBAF; 67). Chemical synthesis of sufficiently pure RNA at scales appropriate for structural studies and therapeutic applications is a still a major challenge. Although some problems like efficient coupling and protecting group removal have been overcome in recent years, an important problem remains: removal of the tetrabutylammonium salts after the 2'-O-silyl deprotection with TBAF. When dealing with hundreds of optical density (OD) measurements of crude oligoribonucleotide the use of TBAF becomes impractical as it demands several parallel chromatographies or use of a large column for desalting purposes and evaporation of large quantities of water. This is cumbersome, requiring unusually large buffer volumes and extensive additional hands-on time, overnight incubation etc. which reduces the quality of the final product. In such cases the use of triethylamine trihydrofluoride $\{Et_3N(HF)_3\}$, may be preferred (68). This reagent generates a less hydrophobic and easier to remove by-product, and is volatile under high vacuum. Most conveniently, the fully deprotected RNA can be precipitated directly from the $Et_3N(HF)_3$ solution with the addition of 1-butanol. Procedures for desilylation with $Et_3N(HF)_3$ are described in the Protocols section. However, it should be noted that the effective use of this reagent is both length and sequence dependent. In addition, care must be taken during work-up due to limited solubility of the products (36, 69) and also during desilylation of chimeric DNA-RNA oligonucleotides. The acidic nature of this reagent may cause some depurination of deoxyadenosine residues.

Analysis and Purification

The fully desalted and fully deprotected RNA oligonucleotide can then be analyzed and purified by reverse phase HPLC, anion-exchange HPLC, PAGE or capillary electrophoresis (analysis only). The fully deprotected oligoribonucleotide is labile to RNase degradation, but this is not a problem except when handling less than 0.1 OD unit. Procedures for preventing RNase degradation have been published (20, 70).

Reverse Phase HPLC

High Performance Liquid Chromatography (HPLC) is an efficient method for the analysis and purification of synthetic oligonucleotides (53). There are many reverse phase (RP) HPLC methods available for the evaluation and purification of oligonucleotides (20-22, 69, 71-73). RP-HPLC purification is generally recommended only for oligoribonucleotides less than 20 bases long provided they give sharp peaks. RNA oligonucleotides with 30 or more bases, or those that form stable secondary structures, generally give broad peaks on reverse phase HPLC analysis. Retention times for DMT-Off RNA oligonucleotides is sequence and length dependent.

Anion Exchange HPLC

Ion exchange columns can also be used for analysis and purification of synthetic oligoribonucleotides. Separation is based upon a variety of factors including the total number of negative charges, total mass, as well as the overall conformation and size of the oligonucleotide in solution. Separation of the crude oligonucleotide is accomplished by gradually increasing the ionic strength of the mobile phase. As a result, the longer, more highly charged oligonucleotides elute later than shorter ones. Oligoribonucleotides that form stable secondary structures or hairpins will deviate from their predictable size-dependent elution pattern and may elute as broad peaks. This effect can be minimized, in most cases, by heating the column. The retention time of the oligonucleotide increases by a few minutes at elevated temperatures. Resolution decreases as longer oligonucleotides are separated and this results in peak broadening. Various methods of ion-exchange HPLC for oligoribonucleotides have been described in reference 20. Oligonucleotides purified by ion exchange HPLC must be desalted before further use.

PAGE

Polyacrylamide Gel Electrophoresis (PAGE) is a traditional and common method for analysis and purification of oligonucleotides. Post-synthesis analysis by PAGE is often performed to verify quality of synthesis. It affords a simple method for analysis of the products from a given synthesis. Methods for analysis and purification of oligoribonucleotides are very similar to those of oligodeoxyribonucleotides and have been well described in literature (20, 53). Although high resolution separations of multiple samples can be achieved, PAGE has limitations. Post-electrophoresis visualization of the oligonucleotides often requires the use of radiolabeling or staining techniques. Furthermore, quantitation of the amount of full-length product and of smaller failure sequences is not possible, except with densitometry.

Capillary Electrophoresis

Gel-capillary electrophoresis is a rapidly evolving new tool for the analysis of oligoribonucleotides and can be routinely used as an alternative to traditional electrophoretic techniques. As the demand for oligoribonucleotides of high purity has increased, demands for additional high resolution analytical techniques have also arisen. The method uses narrow-bore capillaries of fused silica tubing and can rapidly resolve complex mixtures of biopolymers in high electrical fields. We have analyzed synthetic oligoribonucleotides by Micro-Gel capillary electrophoresis (Applied Biosystems Model 270A or 270HT). MicroGel capillaries provide very high efficiency and resolving power for the rapid assessment of the purity of synthetic oligonucleotides (74, 75). Some of the advantages include decreased analysis time, reduced sample quantities (generally 0.1 to 0.2 OD units/100µl is required for analysis) and automation. Purity of a chemically synthesized oligonucleotide can be readily assessed in less than 20 minutes. Long oligonucleotides often require higher sample concentrations or increased injection times due to lower product yield and reduced purity. Run times also need to be increased for longer oligonucleotides. Capillary electrophoresis offers resolution comparable to slab PAGE and higher resolution than reverse phase HPLC methods. This method eliminates tedious and hazardous steps involved in slab gel preparation and development. Changes in ionic strength and pH of the buffer or sample solutions have large effects on elution time, resolution and peak size. Oligoribonucleotides can usually be analyzed after all the 2'-O-TBDMS groups have been removed and the RNA completely desalted.

2'-O-Alkyl Oligoribonucleotides

2'-O-Alkyloligoribonucleotides are proving to be useful reagents for a variety of biological experiments (76). They possess high chemical and thermal stability and are resistant to hydrolysis by DNA- or RNA-specific nucleases. The 2'-O-alkylated oligoribonucleotides hybridize specifically and efficiently to complimentary RNA sequences, forming stable duplexes that are not substrates for RNase H mediated cleavage. RNA oligonucleotides modified at the 2'-hydroxyl position on the ribose have been used as primers for reverse transcriptase reactions (77). Syntheses and applications of 2'-O-methyloligoribonucleotides have been reported in detail (78). Sproat and coworkers have also reported the syntheses of 2'-O-allyloligoribonucleotides (79). These have been shown to increase the level of specific binding to targeted sequences compared with 2'-O-methyl probes of identical sequence (80, 81). 2'-O-Allyl-oligoribonucleotides are ideal for antisense experiments, RNA processing studies, *in situ* localization of specific RNAs and affinity chromatography of RNA-protein complexes. In combination with reporter groups such as biotin, fluorescent dyes, etc. the 2'-O-alkyloligoribonucleotides could play an important role in studies relating to RNA structure and function.

Conclusion and Future Trends

Oligoribonucleotides can today be synthesized efficiently on any automated nucleic acid synthesizer using commercially available phosphoramidites and supports. RNA phosphoramidites with labile base protection and RNA nucleosides loaded on polystyrene support give high quality product. The decrease in ammonia exposure time significantly reduces degradation and potential base modification. It is likely that for synthesis of longer oligoribonucleotides (> 100 bases) significant improvements in chemistry will be needed. These could range from reducing the steric crowding around the reactive phosphoramidite moiety to a novel protecting group for 2'-hydroxyl. Higher coupling yields may be achieved through extremely pure reactive phosphoramidites and better activating agents. It would also be very useful to purify the 2'-O-protected oligoribonucleotides to prevent the ribonuclease degradation during purification of the final product. In addition to large scale RNA preparations, chemical synthesis on the synthesizers permits the incorporation of modifications. It is necessary to develop new reliable techniques for purification of large quantities of synthetic oligoribonucleotides. Providing easier access to catalytic RNAs will also facilitate the synthesis of specific oligonucleotide analogs. The availability of branched RNAs will allow for probing and structural requirements in the splicing reaction. In summary, efforts geared toward elucidating RNA structures, catalytic mechanisms, biological interactions and development of biomedical applications for synthetic oligoribonucleotides will result in the development of better methods of production and purification of oligoribonucleotides. The emerging field of gene therapy (82, 83), the availability of "designer ribozymes" (84) and the ever-expanding catalytic capacity of RNA (85) offer exciting possibilities with synthesis and investigations into the world of RNA. Continuing progress will depend primarily on new experimental results, as chemists, biochemists and molecular biologists work together to address problems concerning molecular replication, *in vitro* selection RNA methods, RNA editing, ribozyme design and engineering, and RNA-based cellular processes. Finally, the biochemists and molecular biologists need to be convinced about the economic value and the efficiency of chemical synthesis of oligoribonucleotides compared with enzymatic transcription methods.

Protocols

Desilylation with tetrabutylammonium fluoride and desalting

Desilylation

Add a 1M solution of tetrabutylammonium fluoride in THF (less than 6 months old and preferably stored under argon) to the crude RNA oligonucleotide (15 μl / OD^{260} unit) after all the base protecting groups have been removed and solvents evaporated. Vortex thoroughly and stir at ambient temperature for 20-24 hours. Quench the reaction with an equal volume of water and concentrate to approximately one half volume.

Desalting (Size-Exclusion Chromatography)

Load a glass column (0.7 x 20 cm) with a slurry of Sephadex G-25 in sterile deionized water. Allow the slurry to flow through the column until the Sephadex has settled (to about 16 cm). Carefully load the RNA sample dissolved in a minimum volume of water (from *desilylation* step). Add about 10 ml of water on the top after the sample has descended to the Sephadex level. A maximum of 100 ODs can be loaded on this column. Collect 10 x 1 ml fractions. RNA elutes in tubes 2-5. Assay each fraction on a UV spectrometer at 260 nm to determine which tube/s contains the RNA oligonucleotide. Pool the appropriate fractions and evaporate to dryness.

Desilylation with Triethylamine Trihydrofluoride and Precipitation of RNA

Add to the crude oligoribonucleotide, a neat solution of triethylamine trihydrofluoride (10 µl / OD^{260} unit). Vortex thoroughly and stir the solution at ambient temperature for 20-24 hours. Quench the reaction with water (2 µl / OD^{260}). To the above solution add 1-butanol (100 µl / OD^{260}), mix and freeze the solution at -20 °C or lower for about an hour. Centrifuge the tube and decant the butanol to collect the precipitated RNA. The precipitated RNA thus obtained can be dissolved in water and analyzed for its purity.

HPLC Analysis and Purification

Reverse Phase HPLC

An Aquapore RP-300 octylsilyl C-8 column (220 x 4.6 mm; Applied Biosystems) can be used for both analytical and purification purposes. The gradient systems for these purposes are (oligoribonucleotides, synthesized DMT-Off):

Gradient System 1:

Start Time (Min)	%B at start time
0	0
24	20
34	40

Mobile phase : Solvent A: 0.1 M triethylammonium acetate/water (H_2O).
Solvent B: Acetonitrile
Flow Rate : 1.0 ml/min

This cartridge is suitable for analysis of about 0.3 to 1.0 OD^{260} unit of crude oligonucleotide. It appears to perform relatively well in separating DMT-On species from DMT-Off. This column may also be used for purifications up to 15 ODs, with some loss in resolution. A larger column (250 x 7 mm) gives better separation with larger loading. Oligoribonucleotides purified by RP-HPLC should then be desalted by size-exclusion chromatography.

Anion Exchange HPLC

A NucleoPac PA-100 ion exchange column (250 x 4 mm: analytical; 250 x 9 mm: semi-preparative, Dionex Corporation) is useful for both analysis and purification purposes (86, 87). The adsorbent is a polymeric anion exchanger, stable to high pH, and high salt and organic concentrations. The gradient system employs lithium perchlorate ($LiClO_4$) in the mobile phase (oligoribonucleotides, synthesized DMT-Off) and the RNA obtained is in lithium salt form. For some experiments, the lithium salt form of RNA may not be suitable and post-purification cation exchange will be required.

Gradient System 2:

Start Time (Min)	%B at start time
0	0
40	70

Mobile phase : Solvent A: 20 mM $LiClO_4$ + 20 mM Sodium acetate (NaOAc) in H_2O: Acetonitrile (CH_3CN) (9:1) (pH 6.5 with dil. AcOH).
Solvent B: 600 mM $LiClO_4$ + 20 mM NaOAc in H_2O: CH_3CN (9:1) (pH 6.5 with dil. AcOH)

Flow Rate : 1.0 mL/min

The analytical cartridge is suitable for loading about 5 ODs of crude oligoribonucleotide. The semi-preparative column may also be used for purification of up to 50 ODs, with some loss of resolution. An even larger column is available for purifying up to 200 ODs of the oligoribonucleotide. Oligoribonucleotides purified by anion exchange HPLC must be desalted. This can be accomplished either by size exclusion chromatography (described previously) or by precipitation of the oligoribonucleotide with the addition of 1-propanol.

The use of lithium perchlorate for anion-exchange based HPLC purifications allows the RNA to be easily isolated salt free by precipitation directly from the product peak fraction by the addition of 1-propanol. Lithium perchlorate is also much more soluble in organic solvents than other perchlorate salts we have examined and this obviates a final desalting column. A separation method based on using sodium perchlorate in the mobile phase is described in reference 87.

Isolation of Purified RNA by Propanol Precipitation

Collect the product peak in a sterile tube (when using Gradient system 2). Add 4 volumes of 1-propanol. Mix thoroughly and keep the tube at -20 °C for 4-6 hours. Centrifuge at 3000-5000 r.p.m for approximately 10 min. Decant the propanol. Wash the precipitated oligoribonucleotide pellet with 1-propanol and dry.

Acknowledgments

The author is grateful to Alex Andrus, Brian Sproat, Peter Wright, and Dean Tsou for helpful discussions and valuable suggestions during the course of this work. The author thanks Chris Marvel and Charles Hotz for review of the manuscript and Ms. Shashi Vinayak for editorial assistance.

References

1. Milligan, J.F., Matteucci, M.D. and Martin, J.C. 1993. Current concepts in antisense drug design. J. Med. Chem. 36: 1925-1937.
2. Uhlmann, E. and Peyman, A. 1990. Antisense Oligonucleotides: A new therapeutic principle. Chem. Rev. 90: 543-584.

3. Rossi, J.J. 1993. Antisense RNA and Ribozymes. In: Methods: A companion to Methods in Enzymology, J.J. Rossi, ed., Academic Press Inc., San Diego, Vol. 5, p 1-5.
4. Melton, D.A. 1988. Antisense RNA and DNA. Cold Spring Harbor Laboratory Press, Cold Spring Harbor, New York.
5. Edgington, S.M. 1992. Ribozymes: Stop making sense. Bio/Technology 10: 256-262.
6. Ojwang, J.O., Hampel, A., Looney, D.J., Wong-Staal, F. and Rappaport, J. 1992. Inhibition of human immunodeficiency virus type 1 expression by a hairpin ribozyme. Proc. Natl. Acad. Sci. USA. 89 : 10802-10806.
7. Cech, T.R. 1990. Self-splicing of group I introns. Annu. Rev. Biochem. 59: 543-568.
8. Symons, R.H. 1992. Small catalytic RNAs. Annu. Rev. Biochem. 61: 641-671.
9. Piccirilli, J.A., McConnell, Zaug, A.J., Noller, H.F. and Cech, T.R. 1992. Aminoacyl esterase activity of the Tetrahymena Ribozyme. Science. 256: 1420-1424.
10. Noller, H.F., Hoffarth, V. and Zimniak, L. 1992. Unusual resistance of peptidyl transferase to protein extraction procedures. Science. 256: 1416-1419.
11. Noller, H.F. 1993. On the origin of the ribosome: Coevolution of subdomains of tRNA and rRNA. In: The RNA World. R.F. Gesteland and J.F. Atkins, eds. Cold Spring Harbor Laboratory Press, New York. p 137-156.
12. Sarver, N., Zaia, J.A. and Rossi, J.J. 1992. Catalytic RNAs (Ribozymes) A new frontier in biomedical applications. In: AIDS Res. Rev. Koff, W.C., Wong-Staal, F. and Kennedy, R.C. eds. Marcel Dekker, Inc. New York. Vol. 2. p. 259-285.
13. Beaudry, A.A. and Joyce, G.F. 1992. Directed evolution of an RNA enzyme. Science. 257: 635-641.
14. Sullenger, B.A. and Cech, T.R. 1993. Tethering ribozymes to a retroviral packaging signal for destruction of viral RNA. Science. 262: 1566-1569.
15. Edington, B.V., Dixon, R.A. and Nelson, R.S. 1993. Ribozymes: Description and uses. In: Books in soils, plants and the environment: Transgenic Plants : Fundamentals and applications. A. Hiatt, ed. Marcel Dekker, New York. p. 301-323.
16. Turner, D.H., Sugimoto, N. and Freier, S.M. 1988. RNA structure prediction. Annu. Rev. Biophys. Biophys. Chem. 17: 167-192.
17. Ratmeyer, L., Vinayak, R., Zhang, Y., Zon, G. and Wilson, W.D. 1994. Sequence specific thermodynamic and structural properties for DNA.RNA duplexes. Biochem. 33: 5298-5304.
18. Puglisi, J.D., Wyatt, J.R. and Tinoco, I. 1991. RNA pseudoknots. Acc. Chem. Res. 24: 152-158.
19. Gray, M.W. and Cedergren, R. 1993. The new age of RNA. FASEB J. 7: 4-6.
20. Gait, M.J., Pritchard, C. and Slim, G. 1991. Oligoribonucleotide synthesis, In: Oligonucleotides and Analogues, A Practical Approach. F. Eckstein, ed. IRL Press, Oxford, England. p 25-48.
21. Vinayak, R. 1993. Chemical synthesis, analysis and purification of oligoribonucleotides. In: Methods: A companion to Methods in Enzymology. J.J. Rossi, ed., Academic Press Inc., San Diego,Vol. 5. p 7-18.
22. Damha, M.J. and Ogilvie, K.K. 1993. Oligoribonucleotide synthesis. In: Methods in Molecular Biology - Protocols for Oligonucleotides and Analogs. S. Agrawal, ed. Humana Press, Totowa, NJ, Vol.20. p 81-114.

23. Reese, C.B. 1987. The problem of 2'-protection in rapid oligoribobnucleotide synthesis. Nucleosides and Nucleotides 6: 121-129.
24. Reese, C.B. 1989. The chemical synthesis of oligo- and poly- ribonucleotides. Nucleic Acids Mol. Biol. 3: 164-181.
25. Milligan, J.F., Groebe, D.R., Witherell, G.W. and Uhlenbeck, O.C. 1987. Oligoribonucleotide synthesis using T7 RNA polymerase and synthetic DNA templates. Nucleic Acids Res. 15: 8783-8798.
26. Wyatt, J.R., Chastain, M. and Puglisi, J.D. 1991. Synthesis and purification of large amounts of RNA oligonucleotides. BioTechniques. 11: 764-769.
27. Nikonowicz, E.P., Sirr, A., Legault, P., Jucker, F.M., Baer, L.M. and Pardi, A. 1992. Preparation of ^{13}C and ^{15}N labelled RNAs for heteronuclear multi-dimensional NMR studies. Nucleic Acids Res. 20: 4507-4513.
28. Batey, R.T., Inada, M., Kujawinski, E. Puglisi, J.D. and Williamson, J.R. 1992. Preparation of isotopically labeled ribonucleotides for multidimensional NMR spectroscopy of RNA. Nucleic Acids Res. 20: 4515-4523.
29. Beckett, D. and Uhlenbeck, O.C. 1984. Enzymatic synthesis of oligoribonucleotides. In: Oligonucleotide Synthesis, A Practical Approach. M.J. Gait, ed. IRL Press, Oxford, England. p 185-197.
30. Beaucage, S. and Iyer, R.P. 1992. Advances in the synthesis of oligonucleotides by the phosphoramidite approach. Tetrahedron. 48: 2223-2311.
31. Goodchild, J. 1992. Enhancement of ribozyme catalytic activity by a contigious oligodeoxynucleotide (facilitator) and by 2'-O-methylation. Nucleic Acids Res. 20: 4607-4612.
32. Taylor, N.R., Kaplan, B.E., Swiderski, P., Li, H. and Rossi, J.J. 1992. Chimeric DNA-RNA hammerhead ribozymes have enhanced *in vitro* catalytic efficiency and increased stability *in vivo*. Nucleic Acids Res. 20: 4559-4665.
33. Chowrira, B.M. and Burke, J.M. 1992. Extensive phosphorothioate substitution yields highly active nuclease-resistant hairpin ribozymes. Nucleic Acids Res. 20: 2835-2840.
34. Slim, G. and Gait, M.J. 1991. Configurationally defined phosphorothioate-containing oligoribonucleotides in the study of the mechanism of cleavage of hammerhead ribozymes. Nucleic Acids Res. 19: 1183-1188.
35. Seela, F., Mersmann, K., Grasby, J.A. and Gait, M.J. 1993. 7-Deazaadenosine: Oligoribonucleotide building block synthesis and autocatalytic hydrolysis of base-modified hammerhead ribozymes. Helv. Chim. Acta. 76: 1809-1820.
36. Adams, C.J., Murray, J.B., Arnold, J.R.P. and Stockley, P.G. 1994. A convenient synthesis of S-cyanoethyl protected 4-thiouridine and its incorporation into oligoribonucleotides. Tetrahedron Lett.: 765-768.
37. Damha, M. J., Ganeshan, K., Hudson, R.H.E. and Zabarylo, S.V. 1992. Solid-phase synthesis of branched oligoribonucleotides related to messenger RNA splicing intermediates. Nucleic Acids Res. 20: 6565-6573.
38. Wang, S. and Kool, E.T. 1994. Circular RNA oligonucleotides. Synthesis, nucleic acid binding properties, and a comparison with circular DNAs. Nucleic Acids Res. 22: 2326-2333.
39. Beaucage, S.L. and Iyer, R.P. The synthesis of specific ribonucleotides and unrelated phosphorylated biomolecules by the phosphoramidite method. Tetrahedron. 46: 10441-10488.

40. Ohtsuka, E. and Iwai, S. 1987. Chemical synthesis of RNA. In: Synthesis and Applications of DNA and RNA. S.A. Narang, ed. Academic Press, Orlando, FL. p 115-136.
41. Scaringe, S.A., Francklyn, C. and Usman, N. 1990. Chemical synthesis of biologically active oligoribonucleotides using ß-cyanoethyl protected ribonucleoside phosphoramidites. Nucleic Acids Res. 18: 5433-5441.
42. Lyttle, M.H., Wright, P.B., Sinha, N.D., Bain, J.D. and Chamberlin, A.R. 1991. New nucleoside phosphoramidites and coupling protocols for solid phase RNA synthesis. J. Org. Chem. 56: 4608-4615.
43. Wu, T., Ogilvie, K.K. and Pon, R.T. 1988. N-Phenoxyacetylated guanosine and adenosine phosphoramidites in the solid phase synthesis of oligoribonucleotides: Synthesis of a ribozyme sequence. Tetrahedron Lett.: 4249-4252.
44. Chaix, C., Molko, D. and Téoule, R. 1989. The use of base labile protecting groups in oligoribonucleotide synthesis. Tetrahedron Lett.: 71-74.
45. Sinha, N.D., Davis, P., Usman, N., Perez, J., Hodge, R., Kremsky, J. and Casale, R. 1993. Labile exocyclic amine protection of nucleosides in DNA,RNA and oligonucleotide analog synthesis facilitating N-deacylation, minimizing depurination and chain degradation. Biochimie. 75: 13-23.
46. Vinayak, R., Anderson, P., McCollum, C. and Hampel, A. 1992. Chemical synthesis of RNA using fast oligonucleotide deprotection chemistry. Nucleic Acids Res. 20: 1265-1269; US Patent No. 5,281,701.
47. Ott, G., Arnold, L., Smrt, J., Sobkowski, M., Limmer, S., Hofmann, H. and Sprinzl, M. 1994. The chemical synthesis of biochemically active oligoribonucleotides using dimethylaminomethylene protected purine H-phosphonates. Nucleosides and Nucleotides. 13: 1069-1085.
48. Rozners, E., Westman, E. and Stromberg, R. 1994. Evaluation of 2'-O-hydroxyl protection in RNA synthesis using H-phosphonate approach. Nucleic Acids Res. 22: 94-99.
49. Ogilvie, K.K. 1983. The alkylsilyl protecting groups : In particular, the t-butyldimethylsilyl group in nucleoside and nucleotide chemistry. In: Nucleosides, nucleotides and their biological applications. J.L. Rideout, D.W. Henry and L.M. Beacham, eds. Academic Press, Orlando, FL. p 209-256.
50. Wu, T. and Ogilvie, K.K. 1990. A study on the alkylsilyl groups in oligoribonucleotide synthesis. J. Org. Chem. 55: 4717-4724.
51. Reese, C.B., Rao, M.V., Serafinowska, H.T., Thompson, E.A. and Yu P.S. 1991. Studies in the solid phase synthesis of oligo- and poly-ribonucleotides. Nucleosides and Nucleotides. 10: 81-97.
52. Beijer, B., Sulston, I., Sproat, B.S., Rider, P., Lamond, A.I. and Neuner, P. 1990. Synthesis and applications of oligoribonucleotides with selected 2'-O-methylation using the 2'-O-[1-(2-fluorophenyl)-4-methoxypiperidinyl-4-yl] protecting group. Nucleic Acids Res. 18: 5143-5151.
53. Applied Biosystems. 1992. The Evaluation and Isolation of Oligonucleotides.
54. Sinha, N.D., Biernat, J., McManus, J. and Köster, H. 1984. Polymer support oligonucleotide synthesis: XVIII: Use of ß-cyanoethyl-N,N-dialkylamino-/N-morpholino phosphoramidite of deoxynucleosides for synthesis of DNA fragments simplifying deprotection and isolation of final product. Nucleic Acids Res. 12: 4539-4557.

55. Atkinson, T. and Smith, M. 1984. Solid-phase synthesis of oligodeoxyribonucleotides by the phosphite triester method. In: Oligonucleotide Synthesis, A Practical Approach. M.J. Gait, ed. IRL Press, Oxford, England. p 35-81.
56. Usman, N., Ogilvie, K.K., Jiang, M-Y. and Cedergren. R.J. 1987. Automated chemical synthesis of long oligoribonucleotides using 2'-O-silylated ribonucleoside 3'-O-phosphoramidites on a controlled glass pore support: Synthesis of a 43-nucleotide sequence similar to 3'-half molecule of an *Escherichia Coli* formylmethionine tRNA. J. Am. Chem. Soc. 109: 7845-7854.
57. McCollum, C. and Andrus, A. 1991. An optimized polystyrene support for rapid, efficient oligonucleotide synthesis. Tetrahedron Lett.: 4069-4072; US Patent No. 5,047,524.
58. Kumar, P., Ghosh, N., Sadana, K.L., Garg, B.S. and Gupta, K.C. 1993. Improved methods for 3'-O-succinylation of 2'-deoxyribo-and ribonucleosides and their covalent anchoring on polymer supports for oligonucleotide synthesis. Nucleosides and Nucleotides. 12: 565-584.
59. Mullah, K.B., Vinayak, R., Andrus, A. and Scarvie, W. 1993. Efficient automated synthesis of RNA oligonucleotides on polystyrene solid support. RNA Processing Meeting of the RNA Society, May 24-29, 1994, Madison, WI, Abstract 298.
60. Vinayak, R., Ratmeyer, L., Wright, P., Andrus, A. and Wilson, D. 1994. Chemical synthesis of biologically active RNA using labile protecting groups. In: Innovations and perspectives in solid-phase synthesis. R. Epton, ed. Mayflower Worldwide Limited, Birmingham, England. p. 45-50.
61. Sproat, B., Colonna, F., Mullah, B., Tsou, D., Andrus, A. Hampel, A. and Vinayak, R. 1995. An efficient method for the isolation and purification of oligo-ribonucleotides. (Nucleosides and Nucleotides. In press).
62. Applied Biosystems User Bulletin 69. 1992. RNA synthesis using phosphoramidites with labile base protection.
63. Lamond, A.I. and Sproat, B.S. 1994. Isolation and characterization of ribonucleoprotein complexes. In: RNA Processing-A Practical Approach, Volume I. S. J. Higgins and B.D. Hames, eds. IRL Press, Oxford, England. p 103-140.
64. Iyer, R.P., Egan, W., Regan, J.B. and Beaucage, S.L. 1990. 3*H*-1,2-Benzodithiole-3-one 1,1-dioxide as an improved sulfurizing agent in solid phase synthesis of oligodeoxyribonucleoside phosphorothiates. J. Am. Chem. Soc. 112: 1253-1254.
65. Stec, W.J., Uznanski, B., Wilk, A., Hirschbein, B.L., Fearon, K.L. and Bergot, B.J. 1993. Bis(O,O-diisopropoxy phosphinothioyl)disulfide - A highly efficient sulfurizing reagent for cost-effective synthesis of oligo(nucleoside phosphorothiate)s. Tetrahedron Lett.: 5317-5320.
66. Vu, H. and Hirschbein, B. 1991. Internucleotide phosphite sulfurization with tetraethylthiuram disulfide. Phosphorothioate oligonucleotide synthesis via phosphoramidite chemistry. Tetrahedron. Lett. 32: 3005-3008.
67. Hogrefe, R.I., McCaffrey, A.P., Borozdina, L.U., McCampbell, E.S. and Vaghefi, M.M. 1993. Effect of excess water on the desilylation of oligoribonucleotides using tetrabutylammonium fluoride. Nucleic Acids Res. 21: 4739-4741.
68. Gasparutto, D., Livache, T., Bazin, H., Duplaa, A., Guy, A., Khorlin, A., Molko, D., Roget, A. and Teoule, R. 1992. Chemical synthesis of a biologically active natural tRNA with its minor bases. Nucleic Acids Res. 20: 5159-5166.

69. Murray, J.B., Collier, A.K. and Arnold, J.R.P. 1994. A general purification procedure for chemically synthesized oligoribonucleotides. Anal. Biochem. 218: 177-184.
70. Sambrook, J., Fritsch, E.F. and Maniatis, T. 1989. Molecular Cloning: A Laboratory Manual. Cold Spring Harbor Laboratory Press, Cold Spring Harbor, New York.
71. Webster, K.R., Shamoo, Y., Konigsberg, W. and Spicer, E.K. 1991. A rapid method for purification of synthetic oligoribonucleotides. BioTechniques. 11: 658-661.
72. Khare, D. and Orban, J. 1992. Synthesis of backbone deuterium labelled $[r(CGCGAAUUCGCG)]_2$ and HPLC purification of synthetic RNA. Nucleic Acids Res. 20: 5131-5136.
73. Usman, N., Egli, M. and Rich, A. 1992. Large scale chemical synthesis, purification and crystallization of RNA-DNA chimeras. Nucleic Acids Res. 20: 6695-6699.
74. Andrus, A. 1994. Gel-capillary electrophoresis analysis of oligonucleotides. In: Methods in Molecular Biology, Vol. 26: Protocols for Oligonucleotide conjugates. S. Agrawal, ed. Humana Press, Totowa, NJ. p 277-300.
75. Demorest, D. and Dubrow, R. 1991. Factors influencing the resolution and quantitation of oligonucleotide separation by capillary electrophoresis on a gel-filled capillary. J. Chromatography. 559: 43-56.
76. Lamond, A.I. and Sproat, B.S. 1993. Antisense oligonucleotides made of 2'-O-alkylRNA: Their properties and applications in RNA biochemistry. FEBS Lett. 325: 123-127.
77. Johansson, H.E., Sproat, B.S. and Melefors, O. 1993. Reverse transcription using nuclease resistant primers. Nucleic Acids Res. 21: 2275-2276.
78. Sproat, B.S. and Lamond, A.I. 1991. 2'-O-Methyloligoribonucleotide: synthesis and applications. In: Oligonucleotides and Analogues, A Practical Approach. F. Eckstein,ed. IRL Press, Oxford, England. p 49-86.
79. Sproat, B.S. 1993. Synthesis of 2'-O-Alkyloligoribonucleotides. In: Methods in Molecular Biology, Vol. 20: Protocols for Oligonucleotides and Analogs. S. Agrawal, ed. Humana Press, Totowa, NJ. p 115-141.
80. Sproat, B.S., Iribarren, A.M., Garcia, R.G. and Beijer, B. 1991. New synthetic routes to synthons for 2'-O-allyloligoribonucleotide assembly. Nucleic Acids Res. 19: 733-738.
81. Iribarren, A.M., Sproat, B.S., Neuner, P., Sulston, I., Ryder, U. and Lamond, A.I. 1990. 2'-O-Alkyl oligoribonucleotides as antisense probes. Proc. Natl. Acad. Sci. USA. 87: 7747-7751.
82. Anderson, W.F. 1994. Gene Therapy for AIDS. Human Gene Therapy 5: 149-150.
83. Kolberg, R. 1994. New AIDS trials seek an answer in gene therapy. J. NIH Res. 6: 29-31.
84. Long, D. M. and Uhlenbeck, O.C. 1993. Self-cleaving catalytic RNA. FASEB J. 7: 15-24.
85. Prudent, J.R., Uno, T. and Schultz, P.G. 1994. Expanding the scope of RNA catalysis. Science. 264: 1924-1927.
86. Thayer, J.R. and Avdalovic, N. 1993. High resolution anion-exchange chromatography of restriction fragments, plasmids and oligonucleotides (normal and phosphorothioate) at pH 8 and 12.4 using a single eluent system on the NucleoPac PA 100. 13th International symposioum on HPLC of proteins, peptides and polynucleotides, San Francisco.
87. Applied Biosystems User Bulletin 79. 1994. RNA Synthesis: Improved post-synthesis protocols, HPLC analysis, and purification.

88. Mujeeb, A., Kerwin, S.M., Egan, W., Kenyon, G.L. and James, T.L. 1992. A potential gene target in HIV-1: Rationale, selection of a conserved sequence, and determination of NMR distance and torsion angle constraints. Biochem. 31: 9325-9338.

From: *Molecular Biology: Current Innovations and Future Trends.*
ISBN 1-898486-01-8 ©1995 Horizon Scientific Press, Wymondham, U.K.

9

ISOELECTRIC FOCUSING OF PROTEINS BY CAPILLARY ELECTROPHORESIS

Tom Pritchett

Abstract

Isoelectric focusing has become an important technique for protein chemists and molecular biologists in academic research, biotechnology R&D, and biotechnology quality control laboratories. Isoelectric focusing in capillaries (CIEF) is a relatively new application of capillary electrophoresis which offers several advantages over traditional slab gel techniques. These include direct detection by U.V. absorbance (eliminating laborious staining and de-staining), faster single-sample analysis, ease of automation, direct transfer of data to a computerized data station for analysis and storage, and the ability to perform fully quantitative analysis. Resolution is comparable to that achieved using traditional techniques involving carrier ampholytes, but has not yet approached that achieved using immobilized ampholytes.

Introduction

The term capillary electrophoresis (CE) refers to separation techniques involving high electrical fields (often greater than 500 V/cm) applied to narrow bore tubes (20 - 200 μm i.d.) usually made of coated or bare fused silica, but sometimes of quartz or glass (1). CE is a general term encompassing several related techniques with distinct separatory mechanisms. Commonly-used modes of CE include free-solution CE, also known as capillary zone electrophoresis (CZE), capillary gel electrophoresis, which is similar to slab gel electrophoresis, micellar electrokinetic capillary chromatography (MECC or MEKC), and isotachophoresis (ITP), which is often used for sample concentration prior to CZE. Capillary isoelectric focusing (CIEF) in free solution, which is comparable to slab or tube gel IEF using carrier (as opposed to immobilized) ampholytes, is a relatively new application of CE.

This chapter is written with the assumption that the reader has some familiarity with the basics of both CE and IEF. Should this not be true, there are several treatises on these subjects (CE: 2-10; IEF: 11-13). Those wishing to familiarize themselves with the wide variety of commercially-available CE instruments are referred to two recent reviews (14, 15).

History of CIEF

H. Svensson, a student of A. Tiselius, is considered by some to have discovered IEF in the early 1960's (16). Others would argue that the first isoelectric focusing experiments were conducted by Kolin in 1954 (1). The first report of isoelectric focusing in capillaries was by Hjertén and Zhu in 1985 (17). Since that time free-solution CIEF has been reported in several different formats including channels with rectangular cross-sections, PTFE tubing, glass tubes (18), and CIEF in gels (19). The most widely-used format today is that of carrier ampholytes in coated or bare fused silica capillaries, and this will be the subject of the following sections.

The use of CIEF is growing rapidly, judging by the recent appearance of many text chapters (1, 20, 21), reviews (22-25), and research papers (see Applications section, below). This growth may be expected to accelerate now that CIEF kits are available from manufacturers such as Beckman (Fullerton California, USA), BioRad (Richmond, California, USA), and Perkin Elmer (Applied Biosystems division, Foster City, California, USA).

Background, Theory, and Principles of CIEF

Basic Principles

In CIEF a pH gradient is formed in a capillary by amphoteric molecules known as ampholytes (a contraction of amphoteric electrolytes) under the influence of an electrical field. Proteins are also amphoteric molecules and for each protein, as for each ampholyte, there is a pH at which the net charge, and therefore the net mobility of the molecule in the field, is zero. This is known as the molecule's isoelectric point (pI). Consequently, when an electrical field is applied to a capillary containing a protein/ampholyte mixture, a pH gradient is formed in which each molecule has migrated to its pI, and a mixture of proteins (or isoforms of a single protein) are separated according to the characteristic pI of each protein or isoform.

The "focusing" aspect of isoelectric focusing can be illustrated as follows. Consider a capillary, with an applied voltage, containing a protein at its pI in a stationary pH gradient. The most acidic part of the gradient will be toward the anode (+) and the gradient will become increasingly basic in the direction of the cathode (-). If the protein diffuses in the direction of the anode it encounters an increasingly acidic environment, and thus acquires a positive charge causing it to migrate by electrophoresis back towards the cathode until its net charge is again zero. Similarly, a protein diffusing towards the cathode acquires a negative charge and migrates back towards the anode. This resistance to diffusion-generated zone broadening is a primary reason that IEF is such a high-resolution method.

A Typical CIEF Experiment

Following is a brief description of the CIEF procedure used by many laboratories. A detailed protocol for one-step CIEF can be found in a later section (Protocols):

1) The sample and standards (proteins whose pIs are known) are mixed with an ampholyte solution to final concentrations of 1-4% ampholytes and 50 - 200 µg/ml of each protein. The ampholyte solution may also contain a viscosity modifier such as methylcellulose or its hydroxypropyl derivative and a gradient extender such as TEMED (18, 20, 22, 26), which also serves to block the so-called "blind end" of the capillary (that is on the side of the detector where one does not want proteins to focus because they will mobilize away from the detector).

2) The sample/standard/ampholyte mixture is introduced into the capillary, usually by positive pressure or vacuum. CIEF can be performed in bare fused silica capillaries (26, 27), but various types of coated capillaries are chosen by most practitioners (22).

3) The ends of the capillary are placed in a basic solution at the cathode and an acid at the anode. If TEMED has not been added to the ampholyte solution, a solution (e.g. catholyte or anolyte) is introduced into the capillary to block the "blind end".

4) Voltage is applied, and focusing takes place. As focusing occurs, the current decreases logarithmically. This results from increasing depletion of charge carriers between pH zones in the capillary (1, 20). The focusing step typically takes 2 - 5 min.

5) The focused zones are mobilized past the detector window, as discussed below. In most formats, basic proteins elute first. Detection is usually accomplished by absorbance at 280 nm, or above for proteins, such as hemoglobins, with prosthetic groups that absorb at higher wavelengths. Lower wavelengths cannot at present be used, due to the absorbance of the ampholytes. The mobilization step typically takes 3 - 30 min.

6) A plot of migration time verses pI is prepared for the standard proteins, from which the pI of the sample is interpolated.

Mobilization in CIEF

The most commonly-used methods of CIEF differ from established gel-based techniques in that the focused protein zones must be mobilized (e.g. past a U.V. detector) to be visualized. The manner in which mobilization is achieved is a primary distinguishing feature of CIEF methods. Focusing and mobilization can be performed as separate steps (two-step methods), or can take place simultaneously (one step method). In the two-step methods capillaries with essentially no electroosmotic flow are used, and the mobilization techniques are typically electrophoretic mobilization or pressure differential mobilization. One-step methods use reduced electroosmotic flow for mobilization. Additionally, CIEF without mobilization is an intriguing possibility currently in the experimental stage.

Electrophoretic Mobilization

Electrophoretic mobilization, also known as chemical mobilization, was first reported by Hjertén and Zhu (17). Electrophoretic mobilization can be toward either the anode or the cathode. Anodic mobilization occurs if the acid at the anode is replaced with base, or with acid containing 20 - 80 mM sodium chloride (or by simply adding NaCl to the anolyte). Conversely, if the base at the cathode is replaced with acid, or if sodium chloride is added to the catholyte, cathodic mobilization occurs. A kit utilizing this method is available from Bio-Rad Laboratories (Richmond, California, USA). For further information on chemical mobilization see references 1, 16, 18, 20, 21, and 28.

Pressure Differential Mobilization

In this type of mobilization, also known as hydrodynamic mobilization and also first reported by Hjertén and Zhu (17), either positive pressure or negative pressure (vacuum) is used to mobilize the focused zones, while voltage is maintained to avoid zone disruption due to such factors as hydrodynamic displacement and diffusion. Two kits employing pressure differential mobilization are available, from Beckman Instruments (Fullerton, California, USA) and from Perkin Elmer/Applied Biosystems (Foster City, California, USA). For further information on pressure differential mobilization see references 18, 21, 29 and 30.

Electroosmotic Flow Mobilization

Electroosmotic flow (EOF), the bulk flow of buffer ions towards an electrode, is present whenever there is a charge on the capillary wall. Fused silica capillaries have a negative wall charge and the EOF is toward the cathode. It was initially assumed that EOF must be eliminated to perform CIEF, but in 1991 Mazzeo and Krull reported a CIEF technique that used a reduced EOF for mobilization (26). No one-step CIEF kits are currently available, but a detailed procedure for this easy to use and flexible technique is included later in this chapter. For further information on EOF mobilization see references 20, 24, 26, 27 and 31-35.

CIEF Without Mobilization

Several formats have been reported or proposed which obviate mobilization. One might scan the detector past the capillary (3, 20) or pull the capillary through the detector (20-22). In addition, several optical absorption imaging detectors which allow continuous monitoring of the CIEF separation have been reported (36). These approaches have many advantages and could find wide application if problems with the capillary requirements (shape, U.V. transparency) and the hardware can be solved in a commercially-viable manner.

Applications of CIEF

CIEF is finding wide application in basic research, protein pharmaceutical development, and biomedicine, and for the purpose of this discussion applications have been divided among these three areas. These categories are presented with the understanding that much crossover exists. For example glutathione transferase will in all likelihood be of diagnostic interest, as may many of the protein pharmaceutical separations.

In the research arena, many reports of separations of standard proteins such as horse heart cytochrome C (pI 9.6), bovine pancreas chymotrypsinogen A (pI 9.1), lentil lectins (pIs 8.6, 8.4, 8.2), horse heart myoglobin (pIs 7.2 and 6.8), human and bovine carbonic anhydrases (pIs 6.6, 5.9), bovine milk beta-lactoglobulin B (pI 5.1), horse spleen ferritin (pIs 4.2 - 4.5), and others, have been published (25, 26, 28, 31-33, 37). Glutathione transferase isoenzymes were separated by Meacher *et al.* (38). Chen and Wiktorowicz developed a quantitative CIEF method for determination of RNase mutants (39). Mazzeo and Krull have even investigated the use of EOF-driven CIEF for peptide mapping (40).

In the area of protein drugs several applications of CIEF have been reported. Glycoforms of recombinant tissue plasminogen activator were analyzed by Yim (41). Wehr *et al.* monitored the degradation of human growth hormone by CIEF (25). Yowell *et al.* used EOF-driven CIEF to analyze a recombinant granulocyte macrophage colony stimulating factor (GM-CSF) dosage form (34, 42). In addition, many groups have reported CIEF analysis of monoclonal antibodies (MAbs). Figure 1 shows one-step CIEF analysis of a murine monoclonal antibody , anti-carcinoembryonic antigen, using wide and narrow range ampholytes (unpublished data). The antibody isoforms in figure 1 range in pI from 6.9 - 7.2, and differ by about 0.1 pI unit. Figure 2 shows the use of CIEF to analyze a model anti-tumor necrosis factor dosage form (43). CIEF has also been used to analyze humanized anti-Tac MAb (44), and a MAb reactive with a wide range of carcinomas (45).

The first CIEF separations published were for hemoglobin and transferrin, two proteins of medical diagnostic interest (17, 20). Both hydrodynamic and electrophoretic

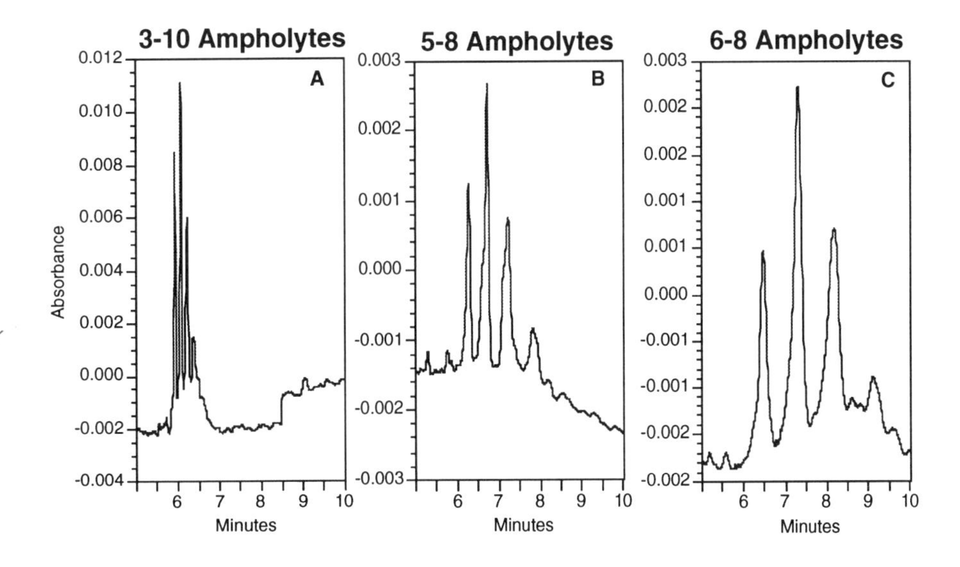

Figure 1. CIEF analysis of anti-CEA MAb. Conditions were as follows: Capillary, eCAP Neutral coated, 37 cm x 50 μm i.d.; Sample concentration, 100 μg/ml; Run temperature, 23 °C; Detection, 280 nm; Applied voltage: **A** and **B**, 250V/cm; **C**, 400V/cm (T. Pritchett, unpublished data).

mobilization were used. Since then several groups have published CIEF analyses of hemoglobins using all three mobilization methods. Zhu *et al.*, using electrophoretic mobilization, separated hemoglobins C, S, F, and A (28), and analyzed abnormal hemoglobins associated with α-thalessemias (46). Hempe demonstrated rapid and quantitative CIEF analysis of normal and abnormal hemoglobin variants using hydrodynamic mobilization (29, 30). Molteni *et al.* accomplished similar analyses using EOF mobilization (27). For a review of the subject of capillary electrophoresis for medical diagnosis see Jellum (47).

Capabilities and Limitations of CIEF

Capabilities

Resolution

Using salt mobilization, Zhu *et al.* resolved hemoglobin variants differing in pI by as little as 0.05 pH units (28) as did Molteni *et al.* using an EOF mobilization method (27). Wehr *et al.* have suggested the possibility of separations on the order of 0.01 pH unit (25).

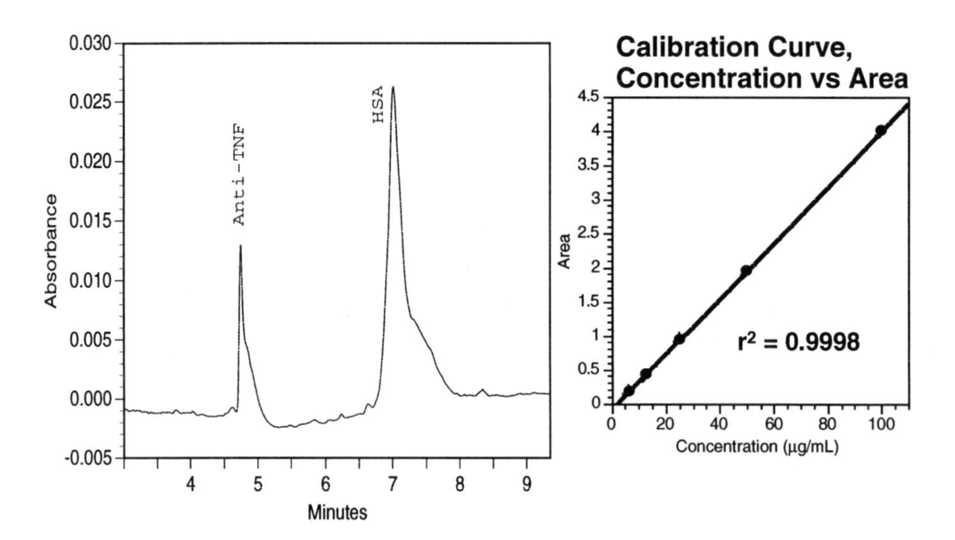

Figure 2. Quantative CIEF analysis of anti-TNF in dosage form with human serum albumin (HSB). Conditions were as follows: Capillary, eCAP Neutral coated, 37 cm x 50 µm i.d.; Sample concentration, 50 µg/ml, Run temperature, 23 °C; Detection, 280 nm; Applied voltage: 300 V/cm. The standard curve was run in the formulation matrix (from reference 43).

Limit of detection

The limit of detection (LOD) for methods which use absorbance at 280nm, the most commonly-used format at present, will depend upon the absorptivity of the protein in question. For large proteins such as monoclonal antibodies, for which $E^{1mg/mL}$ at 280 nm is usually around 1.4, we have found that the LOD is 5 µg/ml (unpublished data). Using the absorption imaging detector discussed above and a scan averaging technique, Wu and Pawliszyn were able to detect 20 µg/ml myoglobin (36).

Reproducibility

Two types of reproducibility are applicable in CIEF. These are migration time (MT) reproducibility and peak area (PA) reproducibility. The latter is of interest when quantitative CIEF is being performed. Reproducibility is reported in terms of percent relative standard deviation (RSD), which is the standard deviation x 100/ the mean.

Using bare fused silica capillaries dynamically coated with polymeric additives Mazzeo and Krull found MT reproducibility to be 2.5% RSD or less (26). Molteni *et al.* reported MT reproducibility of 0.6% - 5% RSD and PA reproducibility of 0.5 - 7.2% RSD for CIEF analysis of hemoglobins (27). Working with a variety of MAbs, the author has achieved migration time reproducibility of 0.5% RSD or less and peak area reproducibility of 1 - 5 % RSD (35).

Linearity

Using EOF mobilization, Mazzeo and Krull have seen correlation coefficients between 0.949 and 0.999 for plots of migration time vs pI, depending upon the standards used (20). Wehr *et al.* (25) and Yowell *et al.* (42) also report a linear relationship between migration time and pI. The author has noted migration time vs pI correlation coefficients of 0.997 (35).

Limitations

Ampholytes

At present the same ampholytes are used for CIEF as for carrier-ampholyte gel IEF. Since these ampholytes are not optimized and quality control tested for CIEF, transparency at 280 nm can vary significantly (both among manufacturers and lot-to-lot). While the appearance of commercial kits (see above) has somewhat alleviated this situation, varieties of broad and narrow range ampholytes specifically designed and tested for CIEF are not yet individually available.

Precipitation

Precipitation of proteins at their pIs is a well-known phenomenon in IEF and is also a problem in CIEF (20). Precipitation is first seen as sharp spikes in the separation pattern, and if severe can cause clogged capillaries. Strategies to minimize precipitation include reducing the protein concentration, reducing the focusing rate, modifying the focusing and mobilization times, and supplementing the sample/ampholyte solution with various additives. Many such additives have been suggested including nonionic surfactants at concentrations (be sure to stay below the CMC) of 0.1 - 2% (e.g., Brij-35, reduced Triton X-100, Nonidet P-40), zwitterionic surfactants (e.g., CHAPS, sulfobetanes), and organic modifiers (e.g., glycerol, 10 - 40 % v/v ethylene glycol; 1, 16, 20). Urea (6 - 8 M) can be used to maintain protein solubility, but at the cost of protein denaturation. When using urea care must be taken (e.g., addition of ion scavengers to stock solutions) to avoid pH dependent carbamylation of the proteins (1, 16).

Detection

The inability to detect at low U.V. wavelengths, due to ampholyte absorbance, limits CIEF to analysis of proteins present in fairly high concentrations, 5-10 μg/ml or greater. Another limitation is that peptides lacking aromatic amino acids can not be detected. Finally, the lack of commercial instrumentation which can monitor or scan the entire capillary makes the choice of a mobilization method necessary for most laboratories.

Matrix Effects

Peak shape and migration time can vary significantly as the total ionic concentration (including protein concentration) in the sample varies. Consequently, the use of external standards or a stored standard curve for pI determination of several samples requires nearly identical ionic and protein concentrations in the standard and sample solutions, which is very difficult to achieve. Therefore, most applications use internal pI standards run with each sample. Consequently, experimentation is required to identify standards which do not co-migrate with samples.

Conclusions and Future Trends

CIEF has advanced greatly since the seminal work of Hjertén in the mid-1980s and now deserves to be considered a routine analytical tool in the protein laboratory. Capillary IEF offers several advantages over the established techniques. CIEF is much less labor intensive than gel IEF, since manual gel handling, and staining/de-staining are eliminated. The direct detection of the capillary format also offers full quantitative capabilities. CIEF, especially the one-step technique detailed here, is also very simple and rapid. Time to first answer can be less than 30 minutes, a decided advantage for industrial applications and when only a few samples need to be analyzed. CIEF is also easily automated for unattended operation, a decided advantage in the increasingly-hectic laboratory of the 1990s.

Much work, however, remains to be done for the full potential of CIEF to be realized. The future, hopefully, will provide elimination or at least amelioration of the limitations discussed above. Namely, future developments should lead to routine methods for preventing protein precipitation, a variety of ampholytes specifically designed for CIEF, improvement of detection, elimination of the need for mobilization, easy procedures for adapting CIEF to a variety of analytical matrices, and the ability to use external standards and stored standard curves. Also, improvements in the stability of capillary coatings at alkaline pH should be forthcoming. Finally, resolution needs to be improved ten fold or so for some applications, especially those of a clinical nature. In this regard, the feasibility of CIEF using immobilized ampholytes seems worthy of investigation.

Protocols: Capillary Isoelectric Focusing Using Electroosmotic Flow Mobilization

The following details a procedure for one-step CIEF in the reversed-polarity mode. This method is, in the author's experience, fast, easy to learn, and adaptable to a wide variety of proteins/matrices and instruments (both commercial and laboratory-built). However, it is difficult to analyze proteins with pIs below 4 with EOF mobilization. For these proteins, pressure differential mobilization methods using one of the commercially-available kits (see above) are suggested.

The example given is that of a monoclonal antibody analyzed on a Beckman P/ACE™ instrument, but the technique is easily adapted to a variety of instruments, provided that the outlet to detector distance is at least 6 cm. For instruments with a very short outlet to detector distance, the method may be run in normal polarity (special instructions given below). Running in normal polarity may also increase resolution in some applications (27).

Special Instructions and Safety Precautions

Capillary electrophoresis involves the use of high-voltage electrical fields. Avoid contact with the instrument while high-voltage is being applied. Always turn off instrument power and unplug the power cord before performing manual polarity reversal.

Making the stock HCl and H_3PO_4 solutions involves handling concentrated acids. Always perform these operations in a fume hood and wear proper safety equipment (goggles, gloves). Always add concentrated acid to water, never the reverse. Have the appropriate spill cleanup and safety wash (eye and body) equipment nearby and readily accessible.

Sodium hydroxide is extremely corrosive. Wear proper safety equipment. Diluting NaOH in water is an exothermic process. The glassware may become hot. Take proper precautions to avoid burns. Have the appropriate spill cleanup and safety wash (eye and body) equipment nearby and readily accessible.

Overview of the Method

Briefly, the method is performed as follows. The CE is set up with a coated capillary to operate in reversed-polarity mode (cathode [-] near the inlet). The catholyte, 20 mM NaOH, is placed at the capillary inlet, and 10 mM phosphoric acid, the anolyte, is placed at the outlet. The capillary is rinsed with 10 mM phosphoric acid, and the entire capillary is filled with the standard and sample protein(s) plus an ampholyte solution containing TEMED (N,N,N',N'-tetramethylethylenediamine) as a gradient extender and HPMC (hydroxypropylmethylcellulose) to provide viscosity and further reduce the EOF. The TEMED concentration is adjusted so that when an electrical field is applied the ampholyte pH gradient forms in the capillary between the detector and the capillary outlet. Upon application of the electrical field (usually 100 - 500 volts per cm) focusing and mobilization back toward the inlet (and past the detector) occur simultaneously, thus avoiding the need for a second mobilization step. Detection is by absorbance at 280 nm, since the polycarboxylic acid-containing ampholytes have significant absorbance at lower UV wavelengths. Detection of proteins at low concentrations (< 10 μg/ml) is often possible because 100 - 200 fold concentration of the protein occurs during focusing.

The isoelectric points of unknown samples are interpolated from a plot of migration time vs pI for standard proteins. The standards are run as internal standards with each unknown or test sample.

Wide-range (pI 3-10) ampholytes are used in this method to maximize the probability of resolving all species present, including possible degradation forms. For maximal resolution, narrow range ampholytes should be used (Figure 1). Many practitioners also make their own custom blends of wide and narrow range ampholytes (sometimes from different suppliers) to fit their analytical needs. Figure 3 shows a typical separation of standards, including an example current trace. Figure 4 shows an example separation of anti-CEA MAb, including internal standards for pI determination.

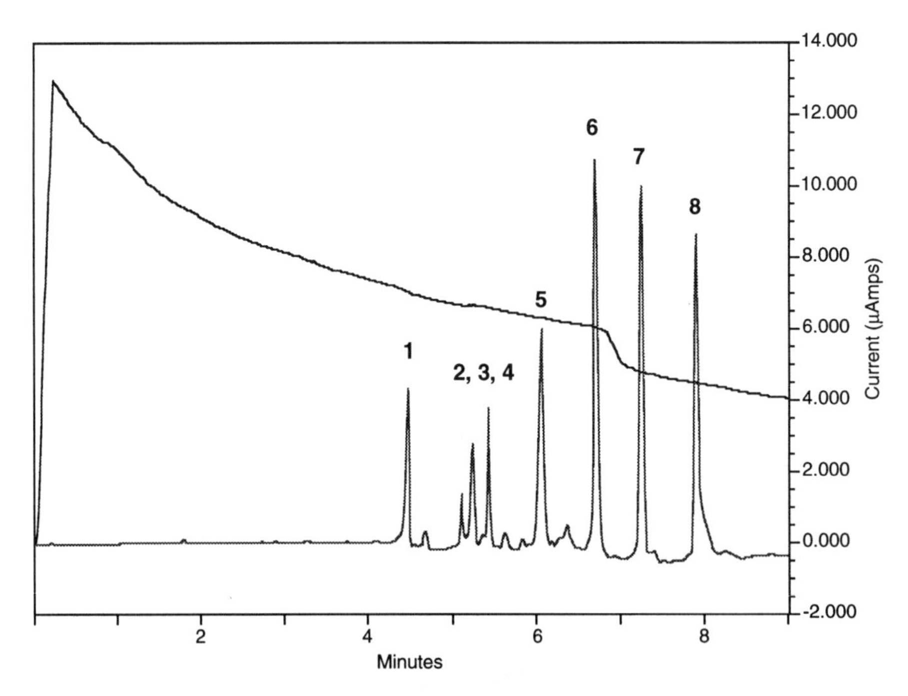

Figure 3. One step CIEF analysis of standards with typical current trace shown. Conditions were as follows: Capillary, eCAP Neutral coated, 37 cm x 50 μm i.d.; Sample concentration, 50 μg/ml of each protein, Run temperature, 23 °C; Detection, 280 nm; Applied voltage: 150 V/cm. Maximum absorbance was 0.045 AU. The identity of the standards are given in figure 4, except for number 6 which is carbonic anhydrase I. (T. Pritchett, unpublished data).

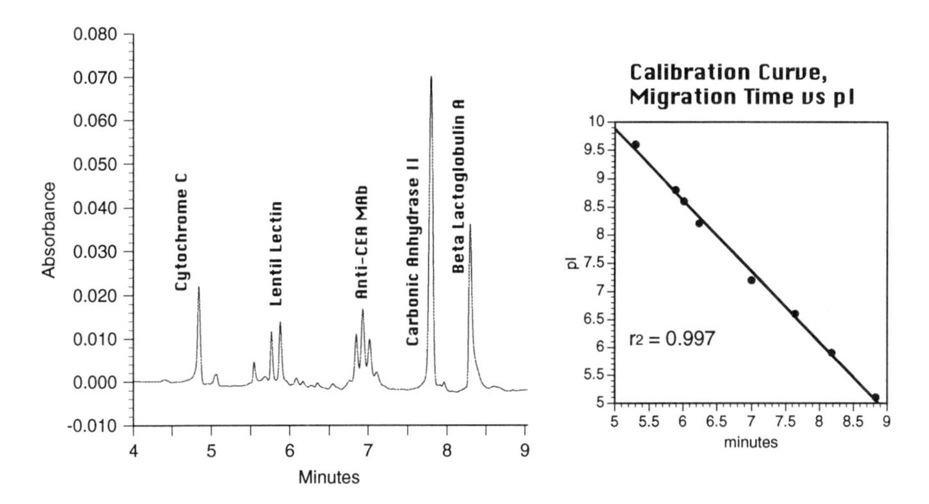

Figure 4. One step CIEF analysis of anti-CEA MAb with internal standards for pI determination. Conditions were as follows: Capillary, eCAP Neutral coated, 37 cm x 50 μm i.d.; Sample concentration, 75 μg/ml each standard, 100 μg/ml of anti-CEA MAb; Run temperature, 23 °C; Detection, 280 nm; Applied voltage: 125 V/cm. A 3-10 ampholyte solution was used, prepared as detailed in this section (from reference 35).

Equipment

Instrumentation

Beckman P/ACE™ or other CE instrument with U.V. detector capable of detection at 280 nm. The method as written works best when the outlet to detector distance is between 6 and 10 cm. IBM compatible computer running System Gold™ or other appropriate CE data capture and analysis software.

Capillary

Several capillaries have been successfully used for this method. The two most popular are the H-150 (C_8) coated capillary from Supelco (Bellefonte, PA. USA) and the eCAP™ Neutral Capillary (Beckman, Fullerton, CA, USA). Other capillary possibilities include the Supelco P-150, and the J&W (Folsom, CA, USA) DB-1 and DB-17 capillaries. Capillary internal diameters (i.d.s) of 50-100 μm are most commonly used. Lengths are usually 25 cm or greater, but shorter lengths have been used.

Other Equipment

Micro centrifuge.
Analytical balance.
Mechanical pipettes and the appropriate disposable tips, covering the range of 1 μl through 1 ml.

Chemicals, Supplies, and Other Disposables

3-10 ampholytes (e.g. Pharmalytes, Sigma P/N P-1522). Ampholytes made by Pharmacia (Piscataway, NJ), BioRad (Richmond, CA), and Serva (Hauppauge, NY, USA) have all been successfully used. Narrow range ampholytes and ampholyte blends may also be used.
N,N,N',N'-tetramethylethylenediamine, (TEMED, Sigma P/N T-9281, or equivalent).
Hydroxypropylmethylcellulose, 4000 cps (HPMC, Sigma P/N H-7509, or equivalent).
Purified water (Milli-Q 18 MΩ reagent water, glass-distilled water, or equivalent).
Hydrochloric acid, ACS reagent (HCl, Sigma P/N H-7020, or equivalent).
Phosphoric acid, ACS reagent, $\geq$ 85% H_3PO_4 (Sigma P/N P-6560, or equivalent).
1.0 N Sodium hydroxide solution (NaOH, Sigma P/N 930-65, or equivalent).
Appropriate tubes (e.g. 0.5 ml polypropylene micro centrifuge tubes), vials (e.g. Beckman P/N 358807), mini-vials (e.g. Beckman P/N 338487) and springs (e.g. Beckman P/N 338488) and/or micro-vials (e.g. Beckman P/N 358819) and springs (e.g. Beckman P/N 358821).
Appropriate vial holders (e.g. Beckman P/N 358818), cooled vial holders (e.g. Beckman P/N 727012), or ambient vial holders for cooled tray (e.g. Beckman P/N 727011).
Vial caps (e.g. Beckman P/N 359079).
0.45 μm filters (Acrodisk™, Gelman Sciences P/N 4184, or equivalent).
10cc plastic syringe with Luer Lock tip (Becton Dickinson P/N 5604 or equivalent).
pI reference standards, for example:

Cytochrome C, pI 9.6 (Horse heart, Sigma P/N C-7752)
Lentil Lectin, pI 8.8, 8.6, 8.2 (Lens Culinaris, Sigma P/N L-9267)
Carbonic Anhydrase II, pI 5.9 (Bovine erythrocyte, Sigma P/N C-6403)
Beta Lactoglobulin A, pI 5.1 (Bovine milk, Sigma P/N L-5137)
(see Applications section above for suggestions for other standards)

Many other proteins are appropriate pI markers. For a comprehensive review see Righetti (48).
Sample: for this example, anti-CEA monoclonal antibody (Pierce, Rockford, IL, P/N 37115).
Appropriate lab ware: glass beakers, 100 ml Class A TC volumetric flasks, 100 ml graduated cylinders, 15 ml polypropylene centrifuge tubes.

Stock Solutions

1% (w/v) HPMC
Add 1g of HPMC (4000 cps) to approximately 90 ml purified water; stir vigorously (2-4 hours, or overnight) to dissolve; add purified water for a final volume of 100 ml.

0.1 M HCl
To approximately 90 ml purified water add 0.86 ml concentrated (11.6 M) HCl; add purified water for a final volume of 100 ml; mix well.

1.0 M H_3PO_4
To approximately 90 ml purified water in a 100 ml Class A TC volumetric flask add 5.525 ml concentrated (18.1 M) H_3PO_4; fill to 100.0 ml; mix well by at least 25 inversions of the flask.

pI Standards
Add sufficient purified water to each standard for a concentration to 4.0 mg/ml; divide each standard into separate 10 μl aliquots, label, date, and store at -80 °C.

Working Solution Preparation

Anolyte and Catholyte

Anode buffer (anolyte), 10 mM phosphoric acid: to approximately 90 ml purified water in a 100 ml Class A TC volumetric flask add 1.0 ml 1 M H_3PO_4; fill to 100.0 ml; mix well by at least 25 inversions of the flask; store tightly capped at ambient temperature for up to 1 month.

Cathode buffer (catholyte), 20 mM sodium hydroxide: to approximately 90 ml purified water in a 100 ml Class A TC volumetric flask add 2.0 ml 1 N NaOH; fill to 100.0 ml; mix well by at least 25 inversions of the flask; store tightly capped at ambient temperature for up to 1 month.

2X CIEF Ampholyte Solution (2 ml):

1) To a 15 ml polypropylene centrifuge tube add:
   - Purified water 0.970 ml
   - 1% HPMC 0.800 ml
   - 3-10 ampholytes 0.200 ml
   - TEMED 0.030 ml (reduce to 0.01 - 0.02 ml and increase water if running in normal polarity, see below).
2) Mix well by vortexing
3) Filter through 0.45 μm Acrodisk™, or other 0.45 μm filter
4) Store at 2-8 °C for up to one week

Example Running Solution for Sample Alone

Sample (e.g. anti-CEA) in ampholyte solution (200 μl final volume; final concentration of sample, 100 μg/ml; this is a good starting concentration for most proteins, but should be optimized for each sample):

1) to a 0.5 ml micro centrifuge tube add:
   - 2X CIEF ampholyte solution 100.0 μl
   - Purified water 80.0 μl
   - Sample at 1 mg/ml 20.0 μl
2) Mix well by vortexing
3) Transfer to a mini-vial
4) Centrifuge at > 10,000 x g for 1-2 min to sediment any particulates and get rid of any bubbles
5) Store at 2-8 °C for up to 48 hours

Example Running Solution of pI Standards

Reference standards in ampholyte solution (200 µL final volume; final concentration of pI standards, 76 µg/ml):

1) to a 0.5 ml micro centrifuge tube add:
   - 2X CIEF ampholyte solution 100.0 µl
   - Purified water 84.8 µl
   - Cytochrome C 3.8 µl
   - Lentil Lectin 3.8 µl
   - Carbonic Anhydrase II 3.8 µl
   - Beta Lactoglobulin A 3.8 µl
2) Mix well by vortexing
3) Transfer to a mini-vial
4) Centrifuge at > 10,000 x g for 1-2 min
5) Store at 2-8 °C for up to 48 hours

Example Running Solution for Sample With Internal Standards for pI Determination

Sample anti-CEA plus pI standards in ampholyte solution (200 µl final volume; final concentrations: pI standards, 76 µg/ml; anti-CEA, 100 µg/ml; ampholytes 2%):

1) to a 0.5 ml micro centrifuge tube add:
   - 2X CIEF ampholyte solution 100.0 µl
   - Purified water 64.8 µl
   - Cytochrome C 3.8 µl
   - Lentil Lectin 3.8 µl
   - Carbonic Anhydrase II 3.8 µl
   - Beta Lactoglobulin A 3.8 µl
   - Anti-CEA at 1 mg/ml 20.0 µl
2) Mix well by vortexing
3) Transfer to a mini-vial
4) Centrifuge at > 10,000 x g for 1-2 min
5) Store at 2-8 °C for up to 48 hours

Analytical Procedure

I. Set up the CE Instrument

A) Place the 20 mM NaOH solution at the capillary inlet, the 10 mM phosphoric acid solution at the outlet. Put the rinse solutions, the sample/ampholyte solution, and a waste vial at appropriate positions.

B) Set the instrument to "reverse-polarity", that is with the cathode (-) at the inlet and the anode (+) at the outlet. For instruments with a very short (shorter than 6 cm) outlet to detector distance, the method may be run in normal polarity (anode at the inlet). In this case, the anolyte should be placed at the inlet, the catholyte should be placed at the outlet, and the amount of TEMED should be decreased to 10 - 20 µl per 2 ml (see ampholyte preparation, above).

II. Program the CE Instrument

A. Time Programming

1) Set capillary temperature to 23 °C.
2) Program a 2 min high-pressure rinse with 10 mM phosphoric acid. Use a separate vial of 10 mM phosphoric acid. Do not use the anolyte.
3) Program a second high-pressure rinse for 4 min with the sample/ampholyte solution.
4) At time 0, program a voltage separation (usually 100 - 500 V/cm) such that the initial current spike is between 15 - 25 μA for neutral to basic proteins and 25 - 35 μA for acidic proteins.
5) At time 12 min, program a 5 min high pressure rinse with purified water.
6) Add a 1 minute rinse with 0.1M HCl as appropriate (e.g. every fifth run) to clean the capillary. Follow the HCl rinse immediately with a 4 minute purified water rinse.

B. Detector Time Programming

1) Set a Data Rate of 2 - 5 Hz.
2) Set Wavelength to 280 nm.
3) Program *Auto Zero* at time 0 min and 1 min.
4) Set instrument to monitor and display current.

III. Perform Runs

Perform runs of ampholyte blank, pI standards, sample, sample with standards, etc.

IV. Plot Migration Time

Plot migration time vs pI for the standards. Interpolate pI(s) of unknown(s) from this curve.

Optimization

To optimize the separation, several parameters may be adjusted. The following are the parameters whose adjustment the author has found most useful:

1) Modify the applied voltage. Lower voltage often gives better resolution for neutral to basic proteins. Higher voltage often gives better peak shape for acidic proteins.
2) Vary the protein concentration.
3) Try different capillaries.
4) Experiment with different ampholytes. Vary the brand, the range, and try blends. Vary the ampholyte concentration.
5) Vary the concentration of TEMED and HPMC. Try other viscosity modifiers (e.g. methylcellulose) and other additives (see above).
6) Try different temperatures.

Troubleshooting

Problem	Possible Cause(s)	Corrective Action(s)
No peaks	Polarity set wrong	Check and correct if necessary
	Initial current spike is too low	Raise voltage until current spike is between 10 - 35 μA
	Detector problems	Check detector setting, lamp
	Capillary window not properly aligned with detector	Check and adjust as necessary
Very low current	Voltage too low	Increase as necessary
	Clogged capillary	Check flow with water. Unclog using syringe and sleeve which fits tightly over capillary (e.g. narrow i.d. Teflon tubing)
Spikes in electropherogram and/or capillaries frequently clogging	Proteins precipitating	Decrease protein concentration, decrease electrical field, use additives (see above under Limitations)
Non-linear plot of migration time vs pI	Deteriorated buffers	Replace
	Inappropriate ampholytes for separation	Try other ampholytes
	Proteins interacting	Incorporate additives which decrease interaction (same as for preventing precipitation, above)
Poor resolution	Ampholyte range too wide	Use narrow range or blend
	Electrical field too high	Decrease
Broad peaks for acidic proteins	Anodic drift	Use capillary with neutral coating; increase anolyte concentration to 20 - 50 mM; increase electrical field to give an initial current spike of 25 - 35 μA

Note: P/ACE ™, System Gold ™ and eCAP ™ are a registered trademarks of Beckman Instruments Inc., Fullerton, CA, USA. Acrodisk ™ is a registered trademarks of Gelman Sciences, Ann Arbor, MI, USA.

References

1. Hjertén, S. 1992. Isoelectric focusing in capillaries. In: Capillary Electrophoresis, Theory and Practice. P.D. Grossman and J.C. Colburn, eds. Academic Press, Inc., San Diego, California. p. 191-214.
2. Landers, J.P. 1994. Handbook of Capillary Electrophoresis. CRC Press, Boca Raton, Florida.
3. Camilleri, P. 1993. Capillary Electrophoresis, Theory and Practice. CRC Press, Boca Raton, Florida.
4. Guzman, N.A. 1993. Capillary Electrophoresis Technology. Marcel Dekker, Inc. New York, New York.
5. Schwartz, H. and Pritchett, T. 1994. Separation of proteins and peptides by capillary electrophoresis: Application to analytical biotechnology. Beckman Instruments, Inc. Fullerton, California. Publication No. 727484
6. Landers, J.P. 1991. High performance capillary electrophoresis of biomolecules. BioEssays. 13: 253-258.
7. Guzman, N.A., Hernandez, L. and Hoebel, B.G. 1989. Capillary electrophoresis: A new era in microseparations. BioPharm. 2: 22-37.
8. Campos, C.C. and Simpson, C.F. 1992. Capillary electrophoresis. J. Chromat. Sci. 30: 53-58.
9. Carchon. H. and Eggermont, E. 1992. Capillary electrophoresis. American Laboratory. January: 67-72.
10. McLaughlin, G.M., Nolan, J.A., Lindahl, J.L., Palmieri. R.H., Anderson, K.W., Morris, S.C., Morrison. J.A. and Bronzert, T.J. 1992. Pharmaceutical drug separations by HPCE: Practical guidelines. J. Liquid Chromatog. 15: 961-1021.
11. Righetti, P.G. 1983. Isoelectric Focusing: Theory, Methodology, and Applications. Elsevier, Amsterdam.
12. Righetti, P.G. and Drysdale, J.W. 1974. Isoelectric focusing in gels. J. Chromatog. 98: 271-321.
13. Righetti, P.G., Gelfi, C. and Gianazza, E. 1986. Conventional isoelectric focusing and immobilized pH gradients. In: Analytical Gel Electrophoreis of Proteins. M.J. Dunn, ed. Wright, Bristol. p. 141-202.
14. Oda, R.P., Spelsberhg, T.C. and Landers, J.P. 1994. Commercial capillary electrophoresis instrumentation. LC•GC. 12: 50-51.
15. Stevenson, R. 1994. A critical review of the development of HPCE instrumentation. J. Cap. Elec. 001: 169-174.
16. Righetti, P.G. and Chiari, M.C. 1993. Conventional isoelectric focusing and immobilized pH gradients: An overview. In: Capillary Electrophoresis Technology. N.A. Guzman, ed. Marcel Dekker, Inc. New York, New York. p. 89-116.
17. Hjertén, S. and Zhu, M-D. 1985. Adaptation of the equipment for high-performance electrophoresis to isoelectric focusing. J. Chromatog. 346: 265-270.
18. Li, S.F.Y. 1992. Capillary Electrophoresis: Principles, Practices, and Applications. Elsevier, Amsterdam. p. 341-347.
19. Dovichi, N.J. 1993. General instrumentation and detection systems. In: Capillary Electrophoresis: Theory and Practice. P. Camilleri, ed. CRC Press, Boca Raton, Florida. p. 25-64.
20. Mazzeo, J.R. and Krull, I.S. 1993. Capillary isoelectric focusing of peptides, proteins, and antibodies. In: Capillary Electrophoresis Technology. N.A. Guzman, ed. Marcel Dekker, Inc. New York, New York. p. 795-818.

21. Kilár, F. 1994. Isoelectric focusing in capillaries. In: Handbook of Capillary Electrophoresis. J.P. Landers, ed. CRC Press, Boca Raton, Florida. p. 95-109.
22. Righetti, P.G. and Gelfi, C. 1994. Isoelectric focusing in capillaries and slab gels: A comparison. J. Cap. Elec. 001:27-35.
23. Schwartz, H. and Pritchett, T. 1994. New approaches to capillary isoelectric focusing of proteins. Bio/Technology. 12: 408-409.
24. Pritchett, T. 1994. One-step capillary isoelectric focusing for rapid analysis of proteins. Genetic Engineering News. June 1: 16.
25. Wehr, T., Zhu, M., Rodriguez, D. and Duncan, K. 1990. High performance isoelectric focusing using capillary electrophoresis instrumentation. Amercian Biotech. Laboratory. September: 22-29.
26. Mazzeo, J.R. and Krull, I.S. 1991. Capillary isoelectric focusing of proteins in uncoated fused-silica capillaries using polymeric additives. Anal. Chem. 63: 2852-2857.
27. Molteni, S., Frischknecht, H. and Thorman, W. 1994. Application of dynamic capillary isoelectric focusing to the analysis of human hemoglobin variants. Electrophoresis. 15: 22-30.
28. Zhu, M., Rodriguez, R. and Wehr, T. 1991. Optimizing separation parameters in capillary isoelectric focusing. J. Chromatog. 559: 479-488.
29. Hempe, J.M. 1994. Hemoglobin analysis by capillary isoelectric focusing. Beckman Application Information Publication No. A-1771-A.
30. Hempe, J.M. and Craver, R.D. 1994. Quantification of hemoglobin variants by capillary isoelectric focusing. J. Clin. Chem. 40: 2288-2295.
31. Mazzeo, J.R. and Krull, I.S. 1992. Improvements in the method developed for performing isoelectric focusing in uncoated capillaries. J. Chromatog. 606: 291-296.
32. Molteni, S. and Thorman, W. 1993. Experimental aspects of capillary isoelectric focusing with electroosmotic zone displacement. J. Chromatog. 638: 187-193.
33. Mazzeo, J.R., Martineau, J.A. and Krull, I.S. 1992. Performance of isoelectric focusing in uncoated and commercially available coated capillaries. Methods. 4: 205-212.
34. Yowell, G.G., Fazio, S.D. and Vivilecchia, R.V. 1993. The analysis of recombinant granulocyte macrophage colony stimulating factor (GM-CSF) by capillary isoelectric focusing. Beckman Application Information Publication No. A-1744.
35. Pritchett, T. 1994. Qualitative and quantitative analysis of monclonal antibodies by one-step capillary isoelectric focusing. Beckman Application Information Publication No. A-1769.
36. Wu, J. and Pawliszyn, J. 1994. Application of capillary isoelectric focusing with adsorption imaging detection to the analysis of proteins. J. Chromatog. B. 657: 327- 332.
37. Chemelík, J. and Thorman, W. 1993. Isoelectric focusing field-flow fractionation and capillary isoelectric focusing with electroosmotic zone displacement. J. Chromatog. 632: 229-234.
38. Meacher, D., Menzel D.B. and Pritchett, T. 1994. Analysis of glutathione S-transferases by capillary isoelectric focusing. Protein Science. 3(Suppl. 1): 142.
39. Chen, S-M. and Wiktorowicz, J.E. 1992. Isoelectric focusing by free solution capillary electrophoresis. Anal. Biochem. 206: 84-90.
40. Mazzeo, J.R. and Krull, I.S. 1993. Peptide mapping using EOF-driven capillary isoelectric focusing. Anal. Biochem. 208: 323-329.

41. Yim, K.W. 1991. Fractionation of the human recombinant tissue plasminogen activator (rtPA) glycoforms by high-peformance capillary zone electrophoresis and capillary isoelectric focusing. J. Chromatog. 559: 401-410.
42. Yowell, G.G., Fazio, S.D. and Vivilecchia, R.V. 1993. Analysis of recombinant granulocyte macrophage colony stimulating factor dosage form by capillary electrophoresis, capillary isoelectric focusing and high performance liquid chromatography. J. Chromatog. 652: 215-224.
43. Pritchett., T. 1994. Quantitative capillary electrophoresis of monoclonal antibodies using the neutral capillary and the SDS 14-200 method development kits. Beckman Application Information Publication No. A-1772-A.
44. Costello, M.A., Woititz, C., De Feo, J., Stremlo, D., Wen, L-F., Palling, D., Iqbal, K. and Guzman, N.A. 1992. Characterization of humanized anti-TAC monoclonal antibody by traditional separation techniques and capillary electrophoresis. J. Liquid Chromatog. 15: 1081-1097.
45. Silverman, C., Komar, M., Shields, K., Diegnan, G. and Adamovics, J. 1992. Separation of the isoforms of a monoclonal antibody by gel isoelectric focusing, high performance liquid chromatography and capillary isoelectric focusing. J. Liquid Chromatog. 15: 207-219.
46. Zhu, M., Wehr, T., Levi, V., Rodriguez, R., Shiffer, K. and Cao, Z.A. 1993. Capillary electrophoresis of abnormal hemoglobins associated with a-thalassemias. J. Chromatog. 652: 119-129.
47. Jellum, E. 1994. Capillary electrophoresis for medical diagnosis. J. Cap. Elec. 001: 97-105.
48. Righetti, P.G. and Caravaggio, T. 1976. Isoelectric points and molecular weights of proteins. J. Chromatog. 127: 1-28.

From: *Molecular Biology: Current Innovations and Future Trends.*
ISBN 1-898486-01-8 ©1995 Horizon Scientific Press, Wymondham, U.K.

10

DNA AND MATRIX ASSISTED LASER DESORPTION IONIZATION MASS SPECTROMETRY

Ivo G. Gut and Stephan Beck

Abstract

Current progress towards the application of matrix assisted laser desorption ionization mass spectrometry for DNA analysis is reviewed. The instrumental set-up is described including recent additions that contribute to improved detection. First applications, such as the analysis of short oligonucleotides for quality control are described. The future of the technique and how it might be used as a substitute for fluorescence automated DNA sequencing are outlined.

Introduction

A large effort is being undertaken to develop full scale automated DNA sequencing. Current DNA sequencing techniques depend on the mass separation of DNA fragments by gel electrophoresis. In the case of conventional Sanger dideoxy sequencing radioactively labelled, single stranded DNA fragments generated from a template DNA using a DNA polymerase are terminated by dideoxy bases (A, T, G, C terminators) in four individual reactions (1). A sequence is read off four parallel running tracks containing the specifically terminated polymerization reactions. Radioactive labels can be substituted with fluorescent ones (2). Protocols using chemiluminescent detection after blotting gels onto a nylon membrane have also been developed with applications in multiplex sequencing (3, 4).

Mass spectrometry is being considered as a replacement of gel based techniques as they are very slow, require a large amount of manual handling and are not easy to automate (i.e. pouring gels, assembly of gels and holders, loading gels). For a recent comprehensive overview of mass spectrometry of biologically relevant material see Burlingame *et al.* (5). With the interest in sequencing entire genomes, throughput in DNA sequencing has turned out to be a bottle neck. The application of mass spectrometry to DNA sequencing has previously been reviewed (6) and since then a slight polarization towards techniques requiring more easily accessible materials has taken place. Matrix assisted laser desorption ionization mass spectrometry (MALDI) has been identified as a likely candidate and most progress has been made here. MALDI finds its major application in the rapid analysis of multicomponent mixtures with good mass accuracy and high sensitivity (7). However the current status is still far from what would be required to make this technology applicable to DNA sequencing. MALDI has successfully been applied to the quality control of short native and modified oligonucleotides (8, 9).

Matrix Assisted Laser Desorption Ionization Time-of-flight Mass Spectrometry (MALDI)

MALDI mass spectrometry (Figure 1) was recently developed by Karas and Hillenkamp (10). It is based on an analyte molecule of interest being ejected into the gas phase and ionized using a matrix and a laser. In most cases the matrix is a low molecular weight organic molecule with optical absorption at the laser wavelength and good proton donor/ acceptor properties. Mechanisms have been discussed by Vertes (11). Interactions taking place in the plume are complex and experiments to determine them require complex instrumentation (12). Large analyte molecules are embedded in matrix molecules at a ratio of roughly 1:1000 although there is considerable variation between different authors. A balance is struck between low concentration and sensitivity of detection of the analyte molecules. A short laser pulse tuned to the optical absorption of the matrix

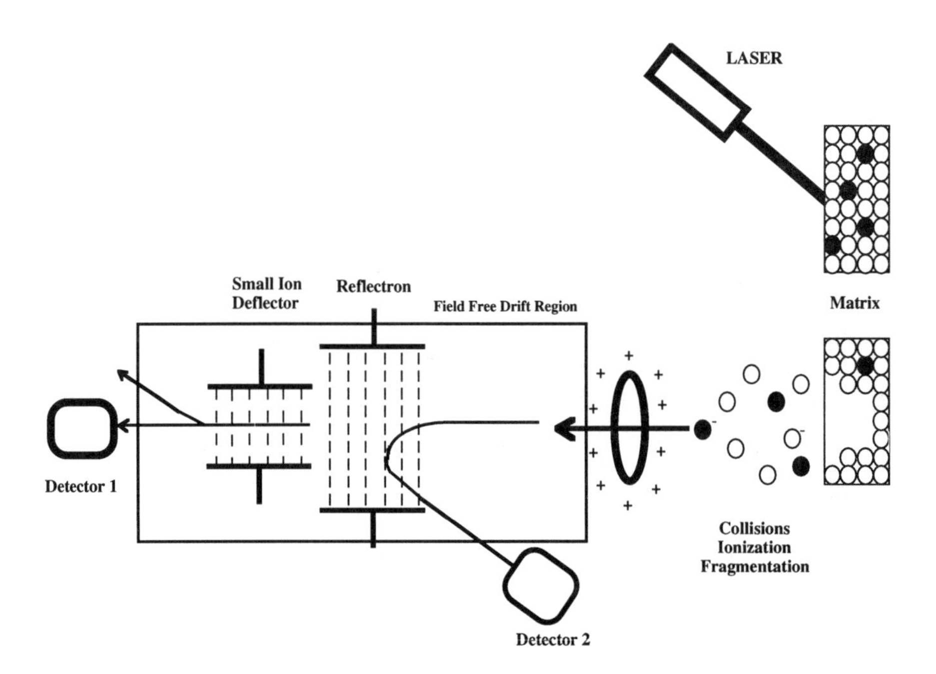

Figure 1. Matrix assisted laser desorption ionization time-of-flight mass spectrometry (MALDI). A laser pulse impinges on a matrix containing analyte molecules. Particles are ejected into the gas phase. Collision with the analyte molecules lead to their ionization. Ionized molecules are accelerated. They can be reflected into detector 2 using the reflectron which results in improved resolution. Detector 1 can be protected against saturation with a small ion deflector that has a pulsed field applied during the early stages of detection. Mainly fragmented matrix material is deflected.

molecules is used for instantaneous vaporization. Vaporization is a central problem of mass spectrometry and defined transfer of molecules from one phase to another is not trivial. The beauty of MALDI is its simplicity of phase transfer. The characteristics of lasers, which are monochromicity and intensity, are a prerequisite for this application. The intensity of a laser beam means that in a very short pulse (1-20 ns) enough energy can be delivered to the matrix for the ablation of sufficient amounts of material for analysis. It also provides a defined starting point for time-of-flight analysis. The monochromicity allows the exclusive excitation of the matrix molecules while analyte molecules are co-ejected into the gas phase without being promoted into an excited state. The analyte molecules remain intact in this process. They are ionized by collisions with matrix molecules. Ions are accelerated in a potential field (several kV) against opposite charge. Through an aperture they enter a field free region in which they drift towards the detector. Fragmentation might take place as a result of collisions after entering the field free region. The momentum of ions is proportional to their mass and charge. Smaller and highly charged species arrive at the detector earlier than large molecules carrying a single charge. Because the matrix molecules are of low molecular

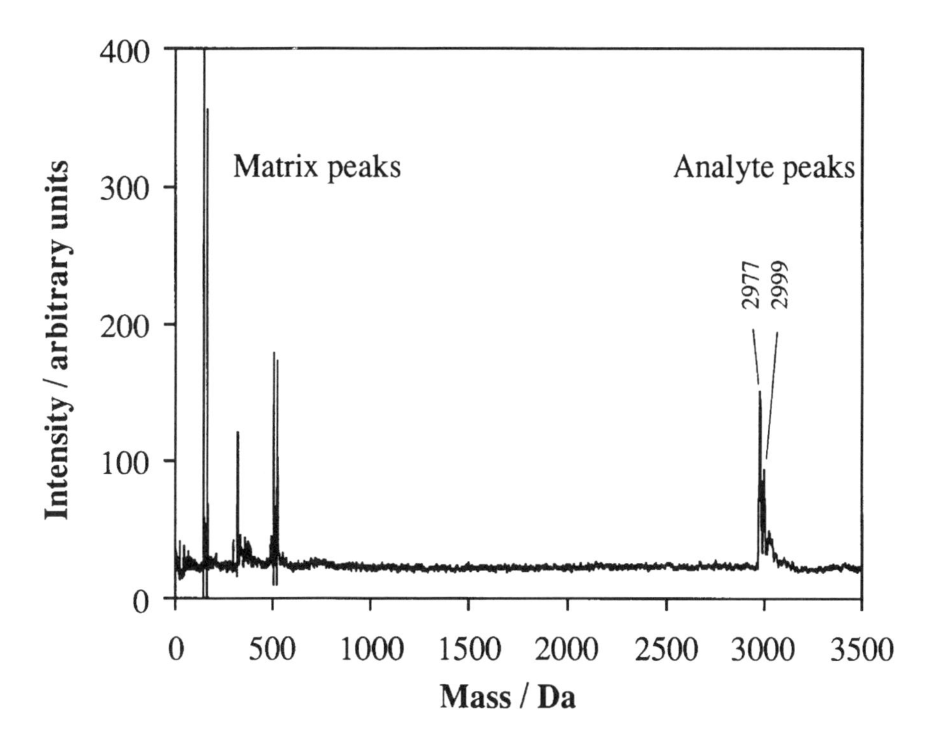

Figure 2. Matrix assisted laser desorption ionization mass spectrum of an oligonucleotide (dT_{10}) recorded in negative ion mode. 2,4,6-Trihydroxyacetophenone was used as matrix. The spectrum was recorded on a FinniganMat instrument which uses a N_2-laser (337 nm) with a pulse duration of 2 ns. Matrix peaks are observed in the lower mass region (< 500 Da). The multiplicity is due to fragmentation and association after laser desorption. dT_{10} is detected at 2977 Da which corresponds to the mass of the free acid - 1 Da. A second, smaller peak is found at 2999 Da. This is the mass of the free acid - 1 Da + 22 Da and corresponds to the free acid with one sodium cation in place of a proton. Further, smaller peaks with larger masses can also be seen. These are higher sodium adducts of dT_{10}.

weight they are observed at the beginning of the analysis (Figure 2). Compared to other mass spectrometers that require the separation of species using the different deviations of charged mass in a magnetic field the time-of-flight analysis only requires the detector to be read out on the time scale of species arriving. The time axis corresponds to the mass/charge ratio.

Lasers and Matrices

Currently the most commonly used lasers are the N_2-laser that emits at 337 nm and the frequency tripled Nd:YAG laser at 355 nm delivering 10-400 mJ/cm^2 in 1-5 ns. The N_2-laser is a gas phase laser. It is reasonably cheap but has a limited lifetime and the output decreases with its use. N_2-lasers have to be replaced when they get too weak, although they usually last more than a year. The solid state Nd:YAG laser is more expensive and requires some maintenance although parts that have to be replaced tend not to be expensive. Solid state lasers can be used for decades if maintained properly. Another advantage of a Nd:YAG laser is a degree of modularity as far as wavelengths are concerned. A Nd:YAG laser with a frequency doubler emits at 532 nm which is a wavelength that has been used for MALDI (13). Nordhoff *et al.* have worked on the feasibility of using infrared lasers for MALDI (14). They used another solid state laser, the Er:YAG that emits at 2.94 μm.

The most popular matrices for the analysis of oligonucleotides are 3-hydroxypicolinic acid, 2,4,6-trihydroxy acetophenone, and 2,5-dihydroxybenzoic acid. Matrix materials characteristically have a strong optical absorption at the laser wavelength they are used in conjunction with. Characteristically they have a number of fairly acidic protons (carboxylic acid and inductively destabilized phenol groups). These labile protons are thought to help protonate the analyte. For native oligonucleotides that carry a lot of negative charges analysis has predominantly been done in negative ion mode. Modified oligonucleotides with uncharged backbones have been analyzed in positive ion mode with sinapinic acid as matrix (9).

In principle MALDI is simple and, in terms of mass spectrometry accuracy, the initial results were crude. However there are devices that have been integrated and found to improve the performance of a MALDI time-of-flight set-up for the analysis of oligonucleotides.

Small Ion Deflectors

Improvement can be achieved by the application of small ion deflectors (15). Small ions are deflected past the detector by means of a pulsed field that is applied perpendicular to the flight path of the ions at the beginning of the detection while mainly matrix fragmentation products in high abundance pass. The information during the first few moments of detection is not of much analytical value as it is not associated with analyte molecules. This leads to an overall decreased number of ions arriving at the detector and an improvement of detected signals is encountered as the detector never approaches saturation.

Reflectrons

The application of a reflectron to the analysis of peptides was shown by Tang *et al.* (16). A potential is applied to an electrostatic mirror. Charged parent and fragment ions are reflected and detected by detector 2 (Figure 1). Neutral fragments resulting from unimolecular decay between target and mirror continue to detector 1. The velocity spread of the reflected spectrum can be corrected and the resolution significantly improved. When applied to oligonucleotides a marked increase in resolution can be observed although data accumulation time does increase as some scanning is involved (17). However when considering DNA sequencing, the resolution is the major critical factor as the distinction between two oligonucleotide sequences differing by one base defines the potential read length of an analysis.

Particle Guides

Nickel wires can be mounted along the centre of the flight tubes (field free region). They are biased at a low voltage so that ions are more efficiently guided down the centre of the flight tube. They are more necessary for long flight tubes (several meters) than short ones as ions tend to diverge from the initial trajectory due to interactions with other ions. These particle guides act like focussing lenses for ions (18).

Suppliers and Manufacturers of Mass Spectrometers

There are several companies manufacturing complete MALDI mass spectrometers that are very user friendly and require very little previous knowledge of the experiment. Brucker, Vestec, Finnigan MAT and Fisons to name just a few. A listing of all suppliers and manufacturers of mass spectrometers was compiled by Lammert (19). There is a constant stream of new machines being introduced to the market and a trend similar to computer pricing can be observed. A machine that was state of the art a year ago and would have cost well over £ 100,000 has been overtaken by an improved model and can now be bought for £ 60,000. Currently the market price for a top class MALDI mass spectrometer with a reflectron and good handling software costs around £ 200,000. When buying a MALDI mass spectrometer one has to outline requirements. The facility of handling a machine might mean that there will not be a lot of room for modulation. The possibility of using a variety of lasers is important if one is interested in studying different matrix materials especially since a different laser/matrix system might emerge that will be the protocol of choice for oligonucleotide analysis. There is quite a significant difference in the software supplied by different companies. This largely determines the facility of data accumulation and the possibilities of data manipulation.

Difference Between the Analysis of Oligonucleotides and Peptides by MALDI

MALDI has been used to determine the masses of large proteins (over 100 KDa) without significant breakup with an accuracy of a few Da (20). DNA is significantly more difficult to analyze by MALDI than proteins. The reasons are obvious when one considers the difference in the nature of oligonucleotides and peptides. Oligonucleotides carry a number of negative charges proportional to the number of bases. Essentially they are salts. Salts have been known to make MALDI analysis difficult as they suppress the matrix. Peptides are made up of mainly neutral amino acids. Out of the 20 natural amino acids only two carry a negative and two carry a positive charge. Although charges are a prerequisit in mass spectrometry the success of an analysis depends on achieving a defined state of ionization. In MALDI the mechanism of ionization is not yet well understood but significant advances are being made (12). A model for the gas phase behaviour of oligonucleotides will be described later.

The Major Steps in the Application of MALDI to the Analysis of Oligonucleotides

Oligothymidylic acids, dT_5 and dT_{10}, were first analyzed by MALDI in 1990 (20, 21). These spectra were recorded in negative ion mode and significant amounts of associated charged species were observed at higher mass than the analyte molecule. Spengler *et al.* reported positive ion spectra of smaller oligonucleotides (3-6 bases) using a system with a laser at 266 nm (22). No significant amount of fragmentation was observed although the background in these experiments was very high. Therefore it would have been hard to observe effects, such as depurination, which occur with a low quantum yield under the high laser intensities used in this work. Nelson *et al.* used a frozen aqueous matrix system on a copper carrier at liquid nitrogen temperature to analyze dA_8 and a 28 base-pair double stranded DNA in positive-ion mode (23, 24). A mixture of oligonucleotides with up to 60 bases was analyzed after desorption of a similar system from a corroded copper surface by Schieltz *et al.* (25). Variation of matrix/laser system (UV versus IR desorption) has been studied extensively by Nordhoff *et al.* (14). Visible wavelength laser desorption was also studied (13, 26). It was found that the wavelength and intensity used determines the amount of fragmentation of the oligonucleotide. The base composition of the oligonucleotide also strongly effects the signal detected (27). Pyrimidine rich oligonucleotides give better signals than purine rich ones. Pure thymine sequences gave by far the best results, whereas pure guanine sequences were difficult to detect. The quality of a signal strongly depends on the matrix (28, 29). Recently it was found that the displacement of sodium by ammonium ions or very stringent desalting protocols improve the definition significantly (8, 30). Pieles *et al.* used their ion-exchange protocol on exonucleolytically digested oligonucleotides and achieved excellent resolution of partially digested oligonucleotide fragments. Their protocol can be used for the sequence analysis of short oligonucleotides and could find an application in quality control for an oligonucleotide synthesis facility.

The molecular weight of a MALDI analyzed oligonucleotide always corresponds to the mass of the free acid minus one proton. This suggests that the counterions are exchanged before the ionization that leads to the acceleration of the species out of the desorption plume. We attribute quite some significance to this observation and will discuss this later. A mock DNA sequence (pooled chemically synthesized oligonucleotides) with up to 40 base oligonucleotides has been analyzed by MALDI (18, 31). The signal strength for equal amounts of larger oligonucleotides in a mixture decreased. It was pointed out that stronger peaks were obtained for larger oligonucleotides when these were analyzed individually. With increasing concentration of analyte larger components of a mixture tend to give weaker signals, which we think is a very important observation. At 40 bases signals became too broad for the distinction between two oligonucleotides differing by one base. However this is a very encouraging development. Recently restriction enzyme digested DNA fragments with lengths up to 100 bases were analysed (32). Although the resolution of this protocol does not yet allow its application, this is a promising step for the analysis of DNA. Kirpekar *et al.* have just presented their very interesting results on the analysis of enzymatically synthesized RNA fragments (33).

Nordhoff *et al.* have been working on the stability of nucleic acids during the matrix assisted laser desorption ionization process (30). It is interesting that single stranded DNA is more prone to breakage of the N-glycosidic bond than RNA with resulting loss of bases. They propose that, in RNA, the 2'-OH group stabilizes the N-glycosidic bond. RNA transcripts with up to 142 nucleotides were shown by MALDI using a reflectron and an IR laser system.

Chemically synthesized methylphosphonate oligonucleotides which have an uncharged backbone show excellent definition in matrix assisted laser desorption ionization in the positive ion mode (9). MALDI analysis of unpurified endproduct could be used for sequence verification as refuse of intermediate steps of the synthesis were detectable. We have been working on the chemical modification of DNA to displace negative charges from the sugar-phosphate backbone. We have developed a protocol that allows the quantitative removal of negative charges. Fully modified oligonucleotides have been analyzed by MALDI in the positive ion mode.

How Could One Picture MALDI of Oligonucleotides?

MALDI of mixtures of peptides usually give well resolved results. For oligonucleotides the results are harder to obtain. In terms of mass spectrometry the major difference between oligonucleotides and peptides is the significantly higher number of charges in oligonucleotides. In a peptide there is not much uniformity to charges as both positively and negatively charged amino acids exist. In peptides with mixed charges a cancellation of charges may even take place. For oligonucleotides all negative charges are identical. Oligonucleotides are always detected with masses corresponding to the free acids although when an oligonucleotide is co-crystallized with a matrix it has a counterion that saturates the negative charge of the sugar-phosphate backbone. Obviously between the vaporization and the detection some quite drastic changes happen to the oligonucleotide which might be responsible for detection difficulties. The separation of the counterions from the oligonucleotide, followed by the saturation of the negative charges with protons from the desorption plume, has to be a concerted process otherwise predominantly higher charged oligonucleotides would be detected or there would be a distribution weighted at higher charge. This has not been observed.

For oligonucleotides M^{-1}, the mass of the oligonucleotide minus one Da carrying one negative charge, is the predominant species (Figure 1). The most likely explanation seems that overall ionization takes place after a majority of the counterions have been displaced by protons in the gas phase. Often Na^+ adduct masses are found (Figure 2). Gas phase affinities of all species of the plume play a significant role. It is found that displacing Na^+ by NH_4^+ leads to a marked improvement of the size of the signals and the resolution. This is understandable from the difference in solution and gas phase acidity of NH_4^+. In solution it is the weaker acid and therefore remains protonated while in gas phase it acts as the stronger acid and protonates the phosphate group. The equilibrium lies strongly on the side of the free acid oligonucleotide and NH_3 and is rapidly reached. If all these processes were 100 % efficient, less of a signal could be detected in negative ion mode as outright negative ionization is far more difficult to achieve than protonation. However that detection in negative ion mode is markedly better indicates a degree of inefficiency of ion exchange leaving a free negative charge for acceleration of the molecule. Peptides are always analyzed in positive ion mode as they are predominantly uncharged in their native state and rely on protonation in gas phase for their analysis by MALDI. Analysis of peptides in negative ion mode hardly yields any signal. When studying oligonucleotides with charged neutral backbones it is found that better signal is obtained in positive ion mode. This suggests that in this case the positive ionization takes place on one of the bases. Excellent resolution of a mixture of three methylphosphonate oligonucleotides was found by Keough *et al.* (9).

Future Trends

It can be expected that significant progress will be made in the development of instrumentation. Already the implementation of deflectors, reflectrons and ion guides has led to a significant improvement of the resolution which is currently easily capable of separating oligonucleotides with 25 bases unambiguously. For larger oligonucleotides a Sanger type approach will have to be used. Four parallel sets of oligonucleotide mixtures each with a specific termination for one of the four bases will have to be generated. The analysis of these four mixtures will have to be carried out individually and the sequence assembled from the set. This would make the analysis independent of the absolute calibration of the system and the larger spacing would require a lower resolution so that measurements could be taken to larger fragments.

Understanding of the gas phase interactions of the matrix material and the analyte molecules should help improve the systems and it is to be expected that some knowledge will be generated here. From current knowledge it seems that there is a certain amount of gas phase interaction that can take place before the MALDI plume exhausts itself. It is likely to be dependent on the amount of matrix material in the plume, the amount of interactions required with the analyte molecules and the time before the plume is cooled beyond an activation barrier or all charged matrix fragments have been removed from the plume by the acceleration potential. The decreased interaction of matrix and analyte required for the analyte to become ionized might improve the resolution as the larger fragments would reach their acceleratable state more rapidly.

Currently for a MALDI analysis, roughly 500 fmoles of an oligonucleotide are required as a specific analyte to matrix ratio has to be maintained and there is a minimum amount of solution that can be deposited. For a single MALDI experiment only a fragment of the deposited matrix is consumed (low fmole to high attomole quantities) while decent signal can be detected. The same matrix spot can be used for virtually hundreds of laser desorptions. The sensitivity level of MALDI is comparable to the DNA found in a single band on a radioactively labelled gel or on a fluorescence sequencer. Usually several spectra are accumulated and then averaged to improve the signal to noise ratio. Development of techniques that would allow accurate deposition of nl's of solution would be desirable for scaling down this technology.

The development of MALDI was published in 1988 (10). Oligonucleotides were first analyzed in 1990 (20,21). In 1993 a mock sequence of 40 bases and a protocol for the analysis of exonucleolytically digested oligonucleotides was described (8,18). Considering the rate of improvement in this field MALDI should soon be applicable to DNA sequencing.

Acknowledgments

We would like to thank Dr. D. Pappin for reading the manuscript.

References

1. Sanger, F., Nicklen, S. and Coulson, A. R. 1977. DNA sequencing with chain-terminating inhibitors. Proc. Natl. Acad. Sci. USA. 74: 5463-5467.
2. Smith, L.M., Fung, S., Hunkapiller, M.W., Hunkapiller, T.J. and Hood, L.E. 1985. The synthesis of oligonucleotides containing an aliphatic amino group at the 5'terminus: Synthesis of fluorescent DNA primers for the use in DNA sequence analysis. Nucleic Acids Res. 13: 2399-2413.
3. Church, G.M. and Kieffer-Higgins, S. 1988. Multiplex DNA sequencing. Science. 240: 185-188.
4. Köster, H., Beck, S., Coull, J.M., Dunne, T., Gildea, B.D., Kissinger, C. and O'Keefe, T. 1991. Oligonucleotide synthesis and multiplex DNA sequencing using chemiluminescent detection. Nucleic Acids Res. 24: 318-321.
5. Burlingame, A.L., Boyd, R.K. and Gaskell, S.J. 1994. Mass spectrometry. Anal. Chem. 66: 634R-683R.
6. Jacobson, K.B., Arlinghaus, H.F., Buchanan, M.V., Chen, C.-H., Glish, G.L., Hettich, R.L. and McLuckey, S.A. 1991. Applications of mass spectrometry to DNA sequencing. GATA. 8: 223-229.
7. Beavis, R.C. and Chait, B.T. 1990. Rapid, sensitive analysis of protein mixtures by mass spectrometry. Proc. Natl. Acad. Sci. USA. 87: 6873-6877.
8. Pieles, U., Zürcher, W., Schär, M. and Moser, H.E. 1993. Matrix-assisted laser desorption ionization time-of-flight mass spectrometry: A powerful tool for the mass and sequence analysis of natural and modified oligonucleotides. Nucleic Acids Res. 21: 3191-3196.
9. Keough, T., Baker, T.R., Dobson, R.L.M., Lacey, M.P., Riley, T.A., Hasselfield, J.A. and Hesselberth, P.E. 1993. Antisense DNA oligonucleotides II: The use of matrix-assisted laser desorption/ionization mass spectrometry for the sequence verification of methylphosphonate oligodeoxyribonucleotides. Rapid Commun. Mass Spectrom. 7: 195-200.

10. Karas, M. and Hillenkamp, F. 1988. Laser desorption ionization of proteins with molecular masses exceeding 10000 daltons. Anal. Chem. 60: 2299-2301.
11. Vertes, A. 1991. Laser desorption of large molecules: Mechanisms and models. In: Methods and Mechanisms for Producing Ions from Large Molecules. K.G. Standing and W. Ens, eds. Plenum Press, New York, USA. p. 275-286.
12. Heise, T.W. and Yeung, E.S. 1994. Spatial and temporal imaging of gas-phase protein and DNA produced by matrix-assisted laser desorption. Anal. Chem. 66: 355-361.
13. Levis, R.J. and Romano, L.J. 1991. Laser vaporization of single-stranded DNA. A study of photoinduced phosphodiester bond scission. J. Am. Chem. Soc. 113: 7802-7803.
14. Nordhoff, E., Kirpekar, F., Karas, M., Cramer, R., Hahner, S., Hillenkamp, F., Kristiansen, K., Roepstorff, P. and Lezius, A. 1994. Comparison of IR- and UV-matrix-assisted laser desorption/ionization mass spectrometry of oligonucleotides. Nucleic Acids Res. 22: 2460-2465.
15. Hedin, A., Westman, A., Hakansson, P. and Sundqvist, B.U.R. 1991. Laser desorption mass spectrometry - some technical and mechanistic aspects. In: Methods and Mechanisms for Producing Ions from Large Molecules. K.G. Standing and W. Ens, eds. Plenum Press, New York, USA. p. 211-219.
16. Tang, X., Ens, W., Poppe-Schriemer, N. and Standing, K.G. 1991. Sensitivity measurements for parent and daughter ions of peptides in a reflecting time-of-flight mass spectrometer. In: Methods and Mechanisms for Producing Ions from Large Molecules. K.G. Standing and W. Ens, eds. Plenum Press, New York, USA. p. 139-144.
17. Wu, K.J., Shaler, T.A. and Becker, C.H. 1994. Time-of-flight mass spectrometry of underivatized single-stranded DNA oligomers by matrix-assisted laser desorption. Anal. Chem. 66: 1637-1645.
18. Fitzgerald, M.C., Zhu, L. and Smith, L.M. 1993. The analysis of mock DNA sequencing reactions using matrix-assisted laser desorption/ionization mass spectrometry. Rapid Commun. Mass Spectrom. 7: 895-897.
19. Lammert, S.A. 1994. 1994 Directory of mass spectrometry manufacturers and suppliers. Rapid Commun. Mass Spectrom. 8: 343-357.
20. Karas, M. and Bahr, U. 1990. Laser desorption ionization mass spectrometry of large biomolecules. Trends Anal. Chem. 9: 321-325.
21. Börnsen, K.O., Schär, M. and Widmer, H.M. 1990. Matrix-assisted laser desorption and ionization mass spectrometry and its applications in chemistry. Chimia. 44: 412-416.
22. Spengler, B., Pan, Y., Cotter, R.J. and Kan, L.-S. 1990. Molecular weight determination of underivatized oligodeoxyribonucleotides by positive-ion matrix assisted ultraviolet laser-desorption mass spectrometry. Rapid Commun. Mass Spectrom. 4: 99-102.
23. Nelson, R.W., Rainbow, M.J., Lohr, D.E. and Williams, P. 1989. Volatilization of high molecular weight DNA by pulsed laser ablation of frozen aqueous solutions. Science. 246: 1585-1587.
24. Nelson, R.W., Thomas, R.M. and Williams, P. 1990. Time-of-flight mass spectrometry of nucleic acids by laser ablation and ionization from a frozen aqueous matrix. Rapid Commun. Mass Spectrom. 4: 348-351.

25. Schieltz, D.M., Chou, C.-W., Lou, C.-W., Thomas, R.M. and Williams, P. 1992. Mass spectrometry of DNA mixtures by laser ablation from frozen aqueous solution. Rapid Commun. Mass. Spectrom. 6: 631-636.
26. Tang, K., Allman, S.L., Jones, R.B. and Chen, C.H. 1992. Comparison of rhodamine dyes as matrices for matrix-assisted laser desorption/ionization mass spectrometry. Org. Mass Spectrom. 27: 1389-1392.
27. Schneider, K. and Chait, B.T. 1993. Matrix-assisted laser desorption mass spectrometry of homopolymer oligodeoxyribonucleotides: Influence of base composition on the mass spectrometric response. Org. Mass Spectrom. 28: 1353-1361.
28. Tang, K., Allman, S.L. and Chen, C.H. 1993. Matrix-assisted laser desorption ionization of oligonucleotides with various matrices. Rapid Commun. Mass Spectrom. 7: 943-948.
29. Currie, G.J. and Yates III, J.R. 1993. Analysis of oligodeoxynucleotides by negative-ion matrix-assisted laser desorption mass spectrometry. J. Am. Soc. Mass Spectrom. 4: 955-963.
30. Nordhoff, E., Cramer, R., Karas, M., Hillenkamp, F., Kirpekar, F., Kristiansen, K. and Roepstorff, P. 1993. Ion stability of nucleic acids in infrared matrix-assisted laser desorption/ionization mass spectrometry. Nucleic Acids Res. 21: 3347-3357.
31. Smith, L.M. 1993. The future of DNA sequencing. Science. 262: 530-532.
32. Tang, K., Allman, S.L., Chen, C.H., Chang, L.Y. and Schell, M. 1994. Matrix-assisted laser desorption/ionization of restriction enzyme-digested DNA. Rapid Commun. Mass Spectrom. 8: 183-186.
33. Kirpekar, F., Nordhoff, E., Kristiansen, K., Roepstorff, P., Lezius, A., Hahner, S., Karas, M. and Hillenkamp, F. 1994. Matrix assisted laser desorption/ionization mass spectrometry of enzymatically synthesized RNA up to 150 kDa. Nucleic Acids Res. 22: 3866-3870.

Index

D

E

F

R

S

T

U

V

Y